Fault Tree Analysis Primer

Clifton A. Ericson II

Copyright © 2011 Clifton A. Ericson II

All rights reserved.

ISBN-10: 1466446102

ISBN-13: 978-1466446106

Printed By CreateSpace Inc., Charleston, NC
CreateSpace.com

CONTENTS

1	Introduction to FTA	1
2	FTA Overview	9
3	FTA Symbols and Terminology	13
4	FTA Process	29
5	FTA Construction	37
6	FTA Construction Examples	47
7	FTA Mathematics	53
8	FTA Rules and Guidelines	71
9	FTA Myths and Criticisms	83
10	FTA Mistakes and Misuses	89
11	FTA Special Topics	93
12	FTA Related Techniques	107
13	FTA Software	113
14	FTA Data	117
15	FTA History	123
A	References	125
B	Example FTA	127
C	About the Author	133
D	Index	135

PREFACE

System safety is an engineering discipline that is applied during system design and development of a product or system to identify and eliminate/mitigate hazards in order to prevent mishaps and accidents. One of the most valuable tools in the system safety toolbox is Fault Tree Analysis (FTA). FTA is also used by other disciplines, such as reliability engineering, systems engineering and accident investigators. FTA is one of the best tools available for detailed root-cause analysis and Probabilistic Risk Assessment (PRA).

A fault tree (FT) is a graphical diagram that uses logic gates to model the various combinations of failures, faults, errors and normal events involved in causing a specified undesired event (UE) to occur. The graphical model can be translated into a mathematical model in order to compute failure probabilities and system importance measures. A FT can model all aspects of a system, including hardware, software, human actions and the environment. FTs are employed to evaluate large, complex and dynamic systems in order to understand and prevent potential safety and reliability problems. Applying the rigorous and structured methodology of FT construction allows the systems analyst to model the unique combinations of fault events that can cause a postulated UE to occur.

In order to design systems that work correctly and safely a system developer must understand and correct the ways by which these systems can go wrong. FTA is highly recommended for detailed root-cause analysis of an UE to determine and understand all of the possible fault combinations that can cause the UE to occur. Once these design errors and flaws have been identified they can be eliminated or mitigated via design safety measures. FTA enjoys a favorable reputation among system safety analysts in all industries utilizing the technique. In some industries it is the only tool that can provide the necessary probability calculations for risk assessments and verification that numerical requirements are being met. Today many commercial computer programs are available for personal desktop computers to assist the FTA analyst in building, editing, mathematically evaluating and printing FTs. FTA has earned its place as a valuable tool for safety, risk assessment, accident investigation, reliability, etc.

The purpose of this book is to provide an introduction to the FTA process and methods; it is an overview as well as a detailed instruction. This book describes the techniques, construction procedures and mathematics of FTA. This book is intended for persons from all industries who are interested in using FTA. It should be very useful to those individuals new to FTA, as well as those practitioners already familiar with the tool. Although FTA has been around for a number of years it is not an outdated tool, it is just as useful and relevant as when it was first developed. It is a proven technique with a highly successful track record.

                                                    Clif Ericson
                                                    Fredericksburg, VA
                                                    December 14, 2011

CHAPTER 1

INTRODUCTION TO FTA

1.1 Background

We live in a world comprised of technological systems, many of which are hazardous. When viewed from an engineering perspective, most aspects of life involve interfacing with systems of one type or another. For example, consider the following types of systems we encounter in daily life: toasters, television sets, home, electrical power, electrical power grid, and hydroelectric power plant. Commercial aircraft are systems that operate within a larger transportation system that includes airport systems, air traffic control systems and a worldwide airspace control systems. The automobile is a system that interfaces with many other systems, such as other vehicles, fuel filling stations, highway systems, bridge systems, traffic control systems, etc.

Systems have become a necessity for modern living, and each system spawns or presents its own unique set of hazards and potential risk because systems have a trait of failing, malfunctioning and/or being erroneously operated. System safety engineering is the discipline and process of developing systems that present reasonable and acceptable mishap risk, for both users and bystanders. Many different tools are used in the system safety discipline to assist in achieving this goal, and Fault Tree Analysis (FTA) is one of the major and most powerful tools available. FTA was invented in the early 1960's to help identify and depict failure combinations in a proposed system design that could potentially lead to a critical nuclear missile system mishap. With the information from the FTA system designers were able to eliminate potential problems and insert barriers and interlocks into the design to preclude a possible future mishap.

In order to design systems that work correctly and safely a system developer needs to understand and correct the ways by which these systems can go wrong. FTA has a favorable reputation among system safety analysts, in all industries, for meeting this need. In some industries it is the only tool that can provide the necessary probability calculations for risk assessments and verification that numerical requirements are being met. Today many commercial computer programs are available for desktop and laptop computers to assist the analyst in building, editing, mathematically evaluating and printing fault trees.

FTA is highly recommended for detailed root-cause analysis of an undesired event (UE) to determine and understand all of the possible fault combinations that can cause the UE to occur. FTA has earned its place as a valuable tool for safety, risk assessment, accident investigation, reliability, etc. There have been criticisms of FTA over the years, but the benefits and strengths of FTA have proven to outweigh the detractor's arguments, and FTA has become an internationally recognized root-cause analysis methodology. It should be noted that FTA is not an outdated tool, but is just as relevant and useful as when it was first developed. It is a proven technique with a highly successful track record.

1.2 What is a Fault Tree?

A fault tree (FT) is a graphical diagram that uses logic gates to model the various combinations of failures, faults, errors and normal events involved in causing a specified UE to occur. The graphical model can be translated into a mathematical model in order to compute failure probabilities and system importance measures. A FT can model all aspects of a system, including hardware, software, human actions and the environment. FTs are employed to evaluate large, complex and dynamic systems, in order to understand and prevent potential safety and reliability problems. Using the rigorous and structured methodology of FT construction, the systems analyst is able to model the unique combinations of fault events that can cause an UE to occur. The UE may be a system hazard, a safety issue of concern, a reliability issue, a problem under exploration or a mishap that is under accident investigation. A FT is a very effective logic diagram methodology for identifying and graphically modeling the various root-causes of a problem or UE.

Figure 1.1 provides a preview of a FT structure using a simplified system example. The system in this case is a chemical processing plant, which has a utility room that contains a gasoline engine powered electrical generator, as well as several electrical motors and electric powered pumps. The UE for this FT is "Fire Destroys System". This simple fuel fire example displays several important aspects of FTA, such as the logic gates, failure events and how they are logically combined together.

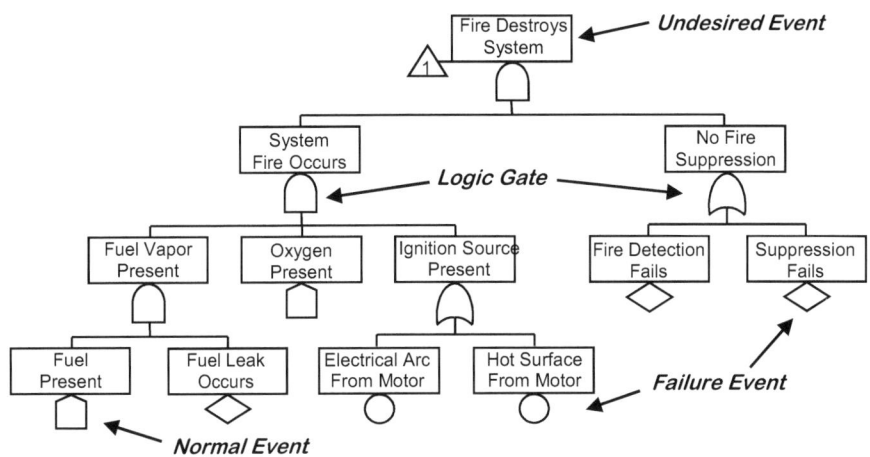

Figure 1.1 – Example Fault Tree

1.3 What is Fault Tree Analysis?

A FTA is a system analysis that uses FTs as the root-cause investigative methodology. FTA determines those combinations of events and conditions that lead to the occurrence of a specified UE. FTA is the analytical process of developing and evaluating a system for the purpose of system improvement. The FTA model logically and graphically brings together the various combinations of possible events, faulty and normal, in a system that can lead to the occurrence of a specified UE. The analysis is deductive in nature, in that it transverses from the general problem to the specific detailed causal factors. FTA develops the logical fault paths through the system that emanate from all of the possible root-causes. The FT of a system graphically depicts the combinations of component failures can yield the top UE. A FTA is sometimes referred to as a Logic Diagram Analysis because of the structured logic involved.

FTA is an analysis methodology for identifying the root-causes of a potential UE and modeling them in the form of a logic diagram. FTA identifies the unique interrelationship of causal factors leading to an UE in order that they can be effectively mitigated through design safety features. The FT diagram logically identifies and displays all of the potential failure modes in a system that can cause the UE to occur. FTA is an excellent tool for anticipating the potential effect of multiple failure combinations. In addition, it is a systems analysis tool because it takes into consideration all parts and aspects of a system, and their interrelationships.

FTA culminates in a logic model of the various combinations of possible events and conditions occurring within a system, which can cause a predefined UE to occur. The UE can be any event (usually objectionable and unwanted) such as a potential accident, mishap, hazardous condition, or undesired failure mode. The graphic FT model exposes the interrelationships of system conditions and events, and their interdependence upon each other.

One of the major strengths of FTA is that it is relatively easy to perform and easy to understand, even by non-experts. In addition, FTA provides useful system design insight that is easily visible from the FT model, as the model reveals all of the possible causal factors for a problem under investigation. The following summarizes the overall strengths and advantages of the FTA technique:

- FTA determines all root-cause events leading to an UE.
- FTA is a structured, rigorous and methodical approach, based on logic.
- FTA models complex system relationships in an understandable manner.
- FTA provides a visual model displaying cause-effect relationships that are very easy to comprehend.
- FTA follows fault paths across system and subsystem boundaries.
- FTA evaluates hardware, software, environmental conditions and human interaction.
- FTA is relatively easy to learn, perform and understand.
- FTA provides a probabilistic risk assessment (PRA).
- FTA can be effectively performed on varying levels of design detail.
- A large portion of the FTA work can be computerized, thereby making the task easier.
- Quality commercial software is available for FTA that runs on desktop and laptop computers.
- FTs can provide design understanding and value despite incomplete system information.
- FTs provide excellent decision making information.
- FTA identifies safety-critical areas in a system.
- FTA is a proven technique with many years of successful use.

- FTA can be used for safety, reliability, unavailability and accident investigation.
- FTA is scientifically sound; it is based on logic theory, probability theory, Boolean algebra and reliability theory.

Although FTA is often classified as a hazard analysis technique, it is primarily used as a root-cause analysis tool to identify and evaluate the causal factors of a specified hazard. FTA is typically not used to identify hazards, but rather to identify the casual factors for an already identified hazard. In addition, it can provide a probabilistic prediction (i.e., risk assessment) of the occurrence of a hazard or UE. The following are different system parameters that can be produced by FTA for system evaluation:

- Cut sets (root-causes) that can cause the UE to occur
- Probability of UE occurrence
- Probability of each cut set occurring
- Probability of system unavailability (downtime)
- Safety criticality of system functions, phases, subsystems and components
- Importance of system components and cut sets and their contribution to the total probability of failure

A FTA is a top down analysis that provides a concise and orderly description of the various combinations of possible fault, failure and normal events within a system that could result in a predefined UE. A FTA essentially starts at the top of the system or subsystem and works downward through the system until it reaches the components failures that lead to the UE. It only includes those failures modes that contribute to the UE. In contrast, the Failure Mode and Effects Analysis (FMEA) is a bottom up analysis that evaluates the ways or modes in which system components can fail individually, and then evaluates the effects of such failures on the system. The FMEA typically starts at the bottom of the system design and evaluates how a component failure will affect the components around it and the overall system. A FTA considers multiple event combinations, whereas the FMEA looks only at single point failures.

1.4 FTA Benefits

FTA is a very powerful and useful systems analysis tool used primarily by system safety, reliability and design engineering. FTA is used to investigate hazards, problems, safety concerns and incidents in an orderly, precise and logical manner. The purpose is to identify and depict the causal factors, conditions and event relationships involved in causing the specified UE under analysis to occur. A quantitative evaluation may be performed, in addition to a qualitative evaluation, to provide a measure of the probability of the occurrence of the top UE and the major faults contributing to the event.

FTA has many different purposes for which it can be utilized, some of which can be summarized as follows:

- To identify the root-causes of an UE during design development in order that they can be eliminated or mitigated early in the design process; a proactive process.
- To determine the specific root-causes of a mishap or incident that has occurred; a reactive process.
- To identify and evaluate diverse multiple failure paths and conditions in systems.
- To calculate the probability of an UE.
- To provide a Probabilistic Risk Assessment (PRA) of a system design.
- To determine high-risk fault paths, and their failure mechanisms, existing in a system design.
- To calculate risk importance measures for system components and fault events in order to determine their criticality.
- To identify design safety deficiencies (subtle or obvious) which have developed in spite of good design requirements and design practices (i.e., design defects).
- To identify common mode and common cause failures that bypass design redundancy.
- To evaluate the efficacy and probability of failure of safety interlocks.
- To establish hazard mitigation methods and evaluate corrective action (design changes).
- To evaluate safety-critical components.
- To verify and demonstrate design compliance with established qualitative and quantitative safety requirements.

FTA can be used to model an entire system, with analysis coverage given to subsystems, assemblies, components, software, procedures, environment and human error. FTA can be conducted at different abstraction levels, such as conceptual design, top-level design, and detailed component design. FTA has been successfully applied to a wide range of systems, such as: missiles, ships, spacecraft, trains, nuclear power plants, aircraft, torpedoes, medical equipment and chemical plants, just to name a few. The technique can be applied to a system very early in design development and thereby identify safety issues early in the design process. Early application of FTA helps system developers to design-in safety from the start rather than having to take corrective action after a test failure or a mishap has occurred.

The completed FT structure can be used to determine the significance of fault events and their probability of occurrence. The validity of action taken to eliminate or control fault events can be enhanced in certain circumstances by performing a quantitative evaluation to evaluate the resulting effect. The numerical evaluation generates three basic measurements for decision making:

1) The probability of occurrence of the top UE
2) The probability and significance of fault events (cut sets) causing the UE
3) The risk significance or importance of components in their contribution to the UE

In most circumstances a qualitative evaluation of the FT will yield useful results that validate a safe design or that show design safety weaknesses that require modification. Careful thought must be given in determining whether to perform a qualitative or a quantitative FTA. The quantitative approach provides more useful results; however, it requires more time and experienced personnel. The quantitative approach also requires gathering component failure rate data for input to the FT.

Since a FT is both a graphic and a logical representation of the root-causes or system faults leading to the UE, it can be used in communicating and supporting decisions to expend resources to mitigate hazards. As such, it provides the required validity in a simple and highly visible form to support decisions of risk acceptability and preventive measure requirements.

The FT process can be applied during any lifecycle phase of a system, from concept to operational usage. FTA should be applied as early in the design process as possible since the earlier necessary design changes are identified and implemented the less they cost. Since FTA provides its greatest strength when applied to the final detailed design, the FT should be updated continuously as the system design progresses.

An important aspect of FTA is the time and cost saving feature of the FTA technique, in that only those system elements that contribute to the occurrence of the UE need to be analyzed. During the analysis, noncontributing elements are ruled out and are, thus, not included in the analysis. This means that a majority of the analysis effort is directed towards the source or sources of the problem area and not on nonproductive areas. Analysis of different UEs would likely result in the analysis of a different set of system elements.

CHAPTER 2

FTA OVERVIEW

2.1 FTA Concept

FTA is a robust, rigorous and structured methodology requiring the application of certain rules of Boolean algebra, logic and probability theory. The FT itself is a logic diagram of all the events (failure modes, human error and normal conditions) that could cause the top UE to occur.

The process of FTA involves analyzing a system, developing the FT model of the system, generating the FT analytical results and evaluating the results for unacceptable design conditions that require possible design modification. FTA is a deductive approach, going from the general to the specific. FTA is an iterative process that considers everything in the system, but only models those components actually involved in the UE. FT construction involves applying a rigorous methodology and rule set.

When the FT model is complete, it is then evaluated to determine the cut sets (CSs) and probability of failure. The CSs are the combination of failure events that can cause the top UE to occur. There are typically many different and unique CSs within a FT; each CS alone will cause the top UE to occur. The FT evaluation provides the necessary information to support risk management decisions.

As shown in Figure 2.1 the theory behind FTA is to start with a postulated top UE (or hazard) and model all of the system faults that can contribute to this top event. The FT model is a direct reflection of the system design, from a failure state viewpoint. In this example, the UE might be "inadvertent missile launch due to system faults".

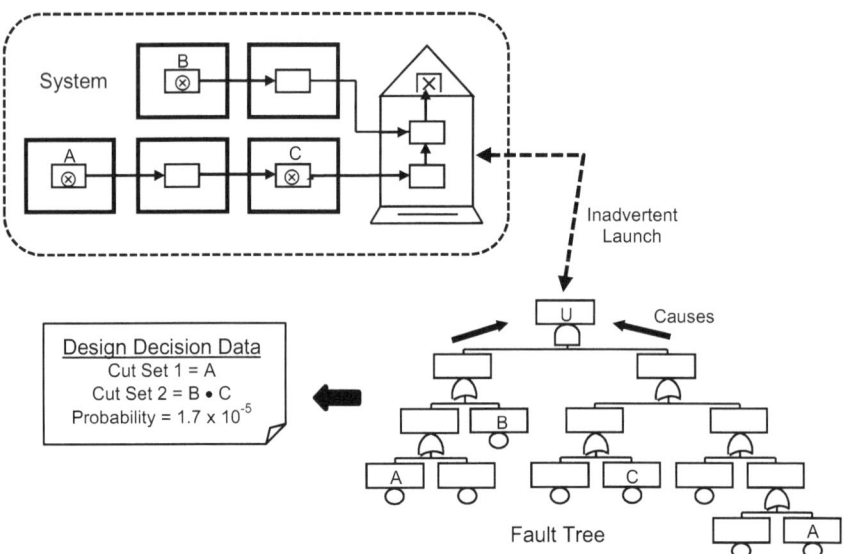

Figure 2.1 – FTA Concept

Figure 2.2 depicts the deductive top-down process, whereby the FT starts at the general level (the system) and traverses down through the system, subsystems and assemblies to the detailed level (components).

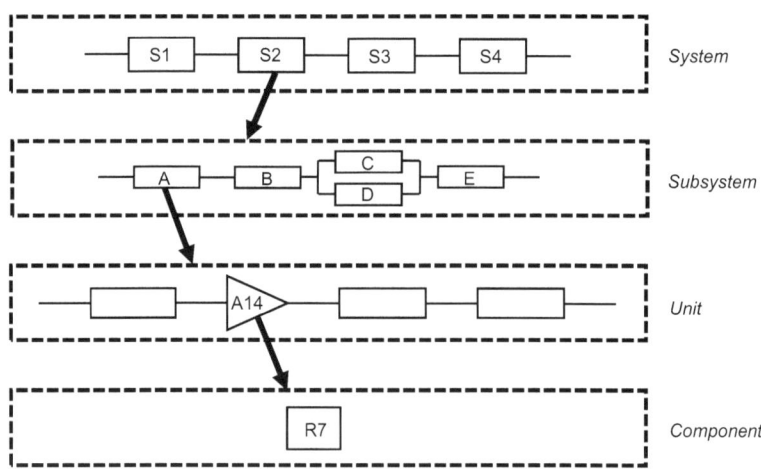

Figure 2.2 – Deductive Approach

FTs are developed in layers, levels and branches using a repetitive analysis process. Figure 2.3 depicts the general FT concept, where the FT expands in layers, with each major layer representing different aspects of the system design. For example, the top FT structure usually models the system functions and phases, the intermediate FT structure models subsystem fault flows and the bottom FT structure models assembly and component fault flows and failures.

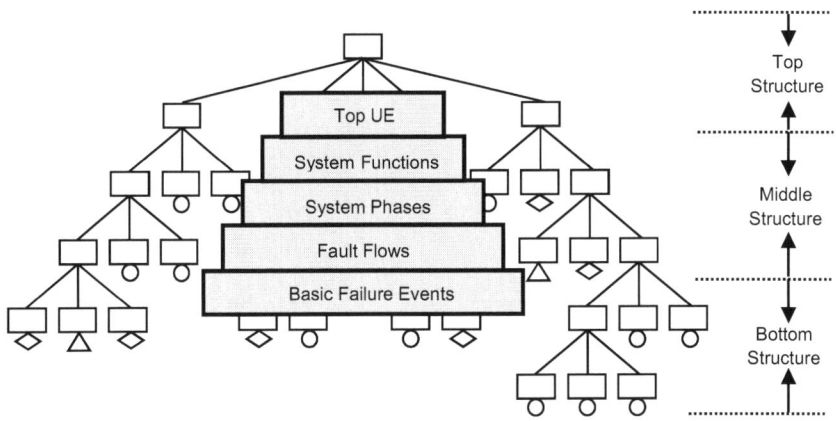

Figure 2.3 – Major Levels of a Fault Tree

FTA should start early in the development program for design evaluation. The goal is to influence design early, before changes are too costly. The FTA should be updated as the design progresses, as shown in Figure 2.4. Each FT update adds more detail to match design detail. Even an early, high-level FT provides useful information regarding the system design.

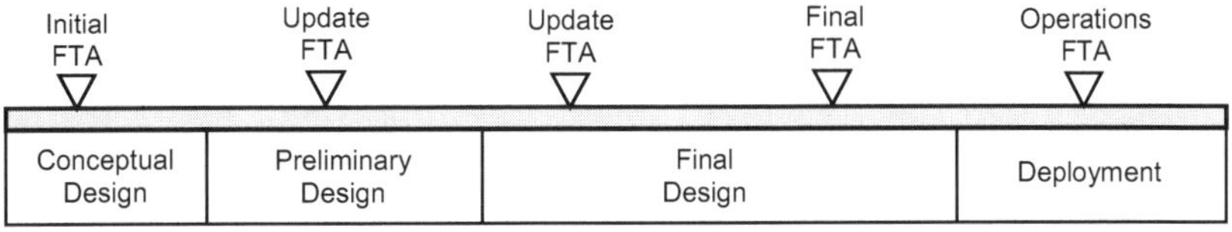

Figure 2.4 – Typical FTA Schedule

Construction of a FT requires considerable design data, safety and hazard data, and reliability data. A risk assessment is the final product of a

FTA, which typically consists of the FT diagram, the generated cut sets, probabilities and importance measures. Figure 2.5 depicts the overall input and output aspects of FTA.

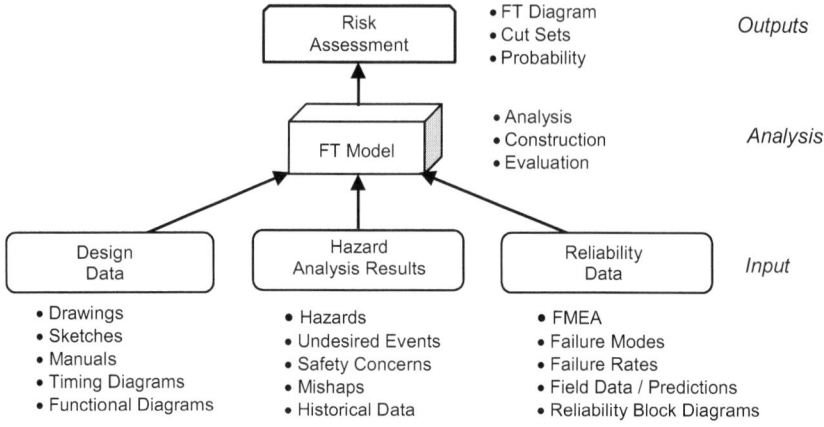

Figure 2.5 – FTA Input-Output Overview

2.2 FT Size Considerations

FTs result in models of various sizes and shapes, depending on the system under investigation and the purpose for the FT. The size of a FT is a critical factor in FTA; the larger the FT the more analysis and bookkeeping effort is required. FTs are typically measured by the number of basic events in the FT, such as the circles, diamonds and house events. Table 2.1 shows the implications of various FT sizes.

Table 2.1 – FT Size Comparisons

Category	Size	Description	Effort
FT Snippet	1 to 2 pages	♦ Presentation; report summary	Small
Small FT	< 100 events	♦ Evaluating a small problem, design option ♦ Example -- aircraft Circuit Breaker Design	Small
Medium FT	100 to 750 events	♦ Detailed analysis of a subsystem ♦ Example -- AWACS Communication System	Medium
Large FT	750 to 2,000 events	♦ Detailed analysis of system ♦ Example -- SRAM Missile System	Large
Huge FT	>2,000 events	♦ Detailed analysis of large system ♦ Example -- Minuteman Multi-Phase FTA	Very Large

CHAPTER 3

FTA SYMBOLS AND TERMINOLOGY

3.1 FT Symbols

FTA is a rigorous analysis technique based upon a special set of symbols, rules, terms and mathematics. The symbols and terms form the basic FT building blocks needed to develop the FT logic diagrams. These building blocks must be thoroughly understood and followed in order to construct and evaluate a FT. The FTA methodology is based upon the following elements:

1) A set of event and gate symbols that provide for a FT logic diagram
2) Boolean algebra that provides for logical combination of gates and events in the FT logic diagram
3) Probability theory that provides for a numerical evaluation of the FT logic diagram
4) Reliability theory for computing event failure probabilities
5) A set of terms and rules that connect the FT components together

Understanding the FTA symbols is paramount to FT construction and evaluation. In addition to the FT symbols, there are a number of special terms and acronyms used in FTA that must be well understood. The following sections describe the FT symbols and terms in detail. The FTA symbols are defined first, followed by the definition of the special FT terms. The detailed construction process is described in chapter 5 and the mathematical aspects are described in chapter 7.

A FT consists of nodes logically inter-linked together in a tree-like structure. These nodes represent fault/failure events and logic gates combined together using the special FT symbols. The node connections form fault paths through the FT structure.

The FT symbols are the primary building blocks of FTA; these symbols fall into the following categories:

1) Basic Events (BEs)
2) Gate Events (GEs)
3) Transfers

Figure 3.1 introduces the various FT symbols, depicting how they appear with text boxes. Basic events consist of the failure modes and normal conditions that can occur in the system. Gate events denote how the basic events are logically combined as *fault* paths are developed through the system. Transfer symbols indicate where a branch or sub-tree is marked for the same usage elsewhere in the FT. The following sections describe these symbols in detail.

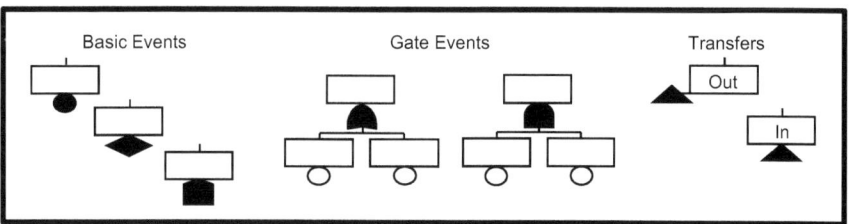

Figure 3.1 – Example FT Symbols

Figure 3.2 shows the standard FT Basic Event (BE) symbols and their definitions. The BE symbols include the circle, diamond and house symbols, which represent failure modes and normally expected events or actions that are possible during system operation. BEs provide the mechanism by which failure rates, exposure times and probabilities enter into the FTA for quantitative evaluations.

Note that the Text Box symbol is nothing more than a placeholder for text for each node or event on the FT. When FTA was first developed, the text was placed directly in the BE symbols (i.e., circle, diamond and house) and the rectangle was only used for Gate Events. With the advent of computer graphics this became cumbersome, so the rectangle was adopted for all nodes in order to standardize the size and amount of text, which makes for an aesthetically better looking FT.

Fault Tree Analysis Primer

Type	Symbol	Description
Text Box	 ▭	The rectangle contains the text for all FT nodes. Text goes in the box, and the node symbol goes below the box.
Primary Failure	 ○	The circle represents the inherent or primary failure mode of a component; a component failure that cannot be further defined in detail.
Secondary Failure	 ◇	The diamond represents a failure that is induced by an external event or failure. It also represents a failure mode that could be developed in more detail if desired.
Normal Event	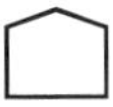 ⌂	The house represents an event or action that is expected to occur as part of normal system operation.

Figure 3.2 – FT Basic Event Symbols

Figure 3.3 shows the Gate Event (GE) symbols and their definitions. These GEs are generally considered the standard FT symbols, although additional gate types have been proposed. Roughly 90% of all FTs can be constructed with just these GE symbols.

Logic Type	Symbol	Description
AND Gate	 ∩	The output occurs only if all of the inputs occur together.
OR Gate	 ∪	The output occurs only if at least one of the inputs occurs.
Priority AND Gate	 ∩─○	The output occurs only if all of the inputs occur together, and in a priority order. The priority statement is contained in the attached condition symbol.
Exclusive OR Gate	 ∪─○	The output occurs if either of the inputs occurs, but not both. The exclusivity statement is contained in the attached condition symbol. Disjoint events.
Inhibit Gate	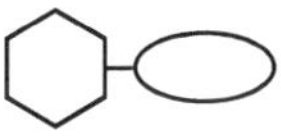 ⬡─○	The output occurs only if the input event occurs and the attached condition is satisfied.

Figure 3.3 – FT Gate Symbols

Figure 3.4 shows alternative forms of the Priority AND (PriAND) and the Exclusive OR (ExclOR) gates. The reason there are alternative forms is that since there is no official FT standard, different FT software vendors have chosen to use different symbols.

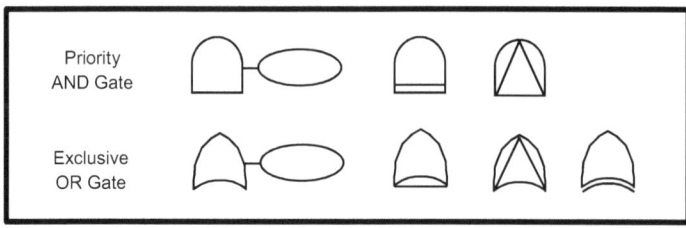

Figure 3.4 –Alternate Symbols for PriAND and ExclOR Gates

Figure 3.5 shows some additional FT symbols that are available for use in FTA. The M of N gate, also known as the Voting Gate, is a shorthand method of denoting m of n combinations rather than drawing them all out long hand.

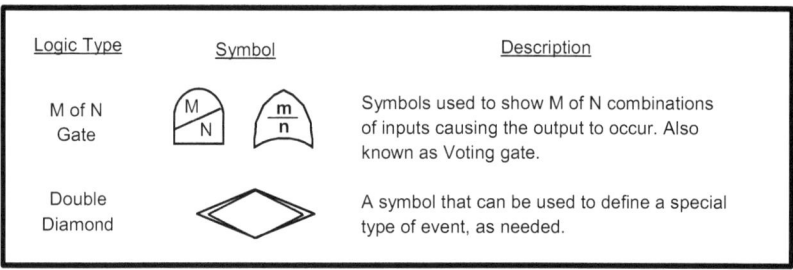

Figure 3.5 – Additional FT Symbols

Figure 3.6 shows the different FT transfer symbols that can be used. The Internal Transfer indicates where a branch or sub-tree is marked for the same usage elsewhere in the current FT. Rather than redrawing duplicate branches they are only drawn once and then additional usages are indicated via the Internal Transfer. Transfer symbols indicate a sub-tree start point and sub-tree insertion points. A FT can have multiple insertion points; these are also known as Multiple Occurring Branches (MOBs) or repeated branches. The transfer symbol is also used to help break-up a FT for printing pages and keeping track of the FT flow.

Fault Tree Analysis Primer

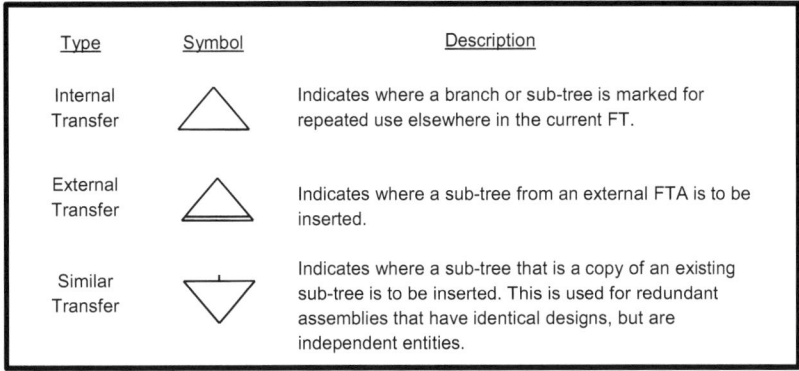

Figure 3.6 – FT Transfer Symbols

3.2 Transfer Symbols and Paging

Figure 3.7 demonstrates the usage of the Transfer symbol within a FT. The Transfer is used for several different purposes, which include:

- To indicate where a branch is used two or more times in the same FT; it is drawn only once rather than repeatedly in several places (Internal Transfer).
- To indicate that there is not enough room on the current printed page for the FT structure; it shows where a branch is placed on a new page (Internal Transfer).
- To indicate an input FT structure from a separate analysis (External Transfer).
- To indicate a FT structure for a separate subsystem that is identical to the current subsystem (Similar Transfer).

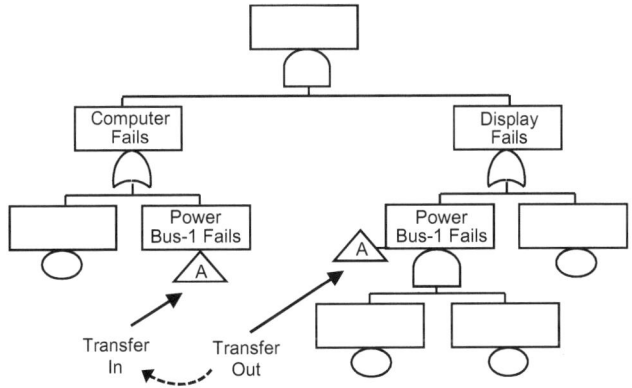

Figure 3.7 – FT Transfer Symbol Example

Figure 3.8 demonstrates the usage of FT Transfer symbol for printed pages and the MOE/MOB concept. This figure shows an example of three printed FT pages. On page 1, the node with a triangle at the bottom with the name A, represents a transfer-in. This means that a duplicate of branch A should also be inserted at this point, but it is drawn somewhere else, page 2 in this case. In this case, Transfer A is not an MOB but merely the transfer of tree to start on a new page, due to lack of space on page 1. Transfer C represents an MOB, as it is intended to be repeated in two different places in the FT. It is defined on page 2, and used again on pages 1 and 3, in this example.

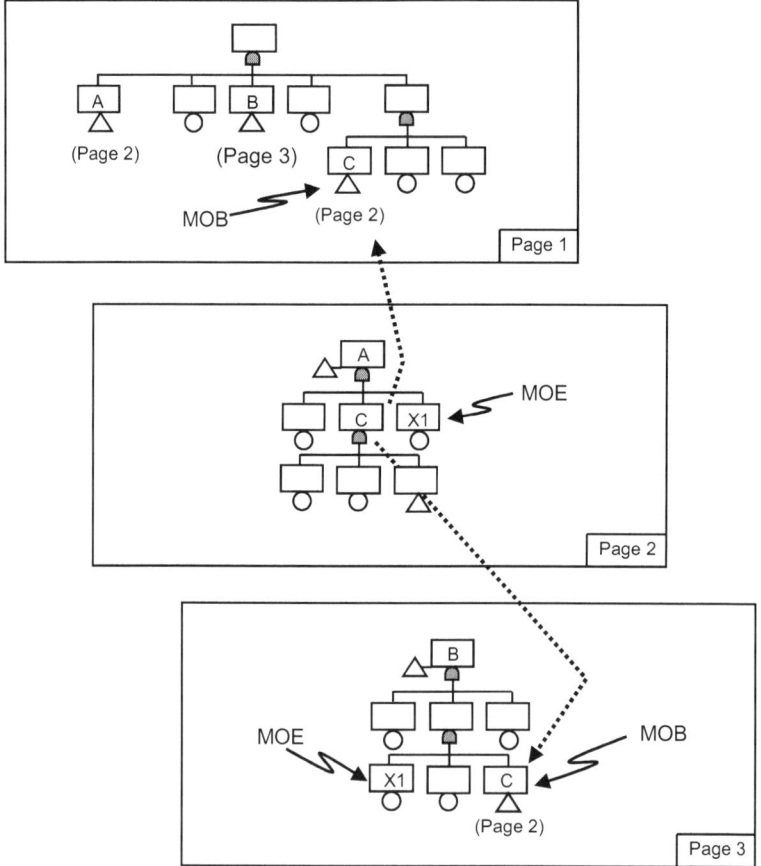

Figure 3.8 – FT Paging Using Transfer Symbols

3.3 M of N Gate

The M of N gate is a shorthand notation for a larger FT structure. Figure 3.9 shows an example of the M of N gate in both the shorthand and full FT structures. Gate output occurs when M of N of the inputs occur. The M of N gate is used to eliminate drawing all the possible combinations in the tree structure. This can be very useful when the full FT structure of all the combinations is very large. A FT computer program should handle the mathematics automatically.

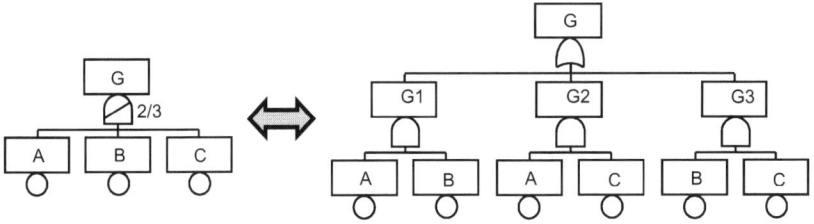

Figure 3.9 – M of N Gate Example

3.4 FTA Terms

The following definitions are provided for terms that are used specifically within the FTA discipline. It is important for the FTA analyst to fully comprehend these terms.

<u>Coherent FT</u>

A coherent FT is a FT that does not contain NOT gates. It should be noted, however, that an Exclusive OR gate contains an implied or hidden NOT.

<u>Coherent System</u>

A system that when in the failed state, cannot be returned to the non-failed state by an additional component failure. Also, when in the non-failed state, the system cannot be made to fail by the repair of a component.

<u>Command Fault</u>

A command failure of an item is when the item is "commanded" to fail, or forced into a fault state by another event. This usually happens when the inadvertent operation of one component affects the operation of another component; one component misleads another. For example, "Light Off" is the command fault for a light when the power to the light fails (the light itself has not actually failed, it has been turned off unintentionally through various possible failures). An analogy for command failure is the falling of dominoes, whereby each domino is caused to fall by the one next to it. A command fault can be the normal operational state, but, at the wrong time, and sometimes it is the lack of the desired normal state when intended. This is the *transition* event to look for in FT construction.

Common Cause Failure (CCF)

A CCF is the failure of failure of two or more components due to a single causal source. CCFs are multiple failures that result from, or are caused by, seemingly unrelated failures or an adverse environment. For example, two or more components may fail due to the single cause of an over temperature environment. The mode of failure for each of the components may not be the same. Common causes may be conditions or events internal within the system or external, from its environment. CCF is an event that simultaneously affects a number of components that were otherwise considered to be independent. CCF failure is a mechanism that defeats or invalidates design redundancy and independence.

Common Mode Failure (CMF)

A CMF is the failure of two or more components due to a single causal source; however, the components fail in the same mode. For example two or more components such as switches may fail in a single mode such as an open circuit. The reason for the failure may not be the same. A manufacturing error, or a material failure, may result in the CMF of two different, but identical, components manufactured in the same batch. CMF is an event that simultaneously affects a number of components that were otherwise considered to be independent. CMF failure is a mechanism that bypasses or invalidates design redundancy and independence.

Critical Path

A FT critical path refers to the CS having the highest probability. This is a major CS that drives the FT top UE probability. It is referred to as a critical path because it is a path of events that if changed, will make most dramatic system improvement in terms of probability reduction.

Cut Set (CS)

A CS is a unique set of events (e.g., failures) that together cause the FT top UE to occur. CSs establish unique paths leading to the UE. A FT can have a few cut sets or thousands, depending on the size and complexity of the FT. Each CS is an independent root-cause for the top UE. There are several categories of cut sets:

- Minimal cut set – a CS that cannot be further reduced and still produce occurrence of the top UE.
- Super cut set – A non-minimal CS; an element can be removed and the FT top event will still occur. These are eliminated in the CS reduction process.
- Duplicate cut set – a repeated CS that occurs due to the unique structure of the system. These are eliminated in the CS reduction process.

Cut Set Order

The CS order is the number of items in a CS. A 1-order CS is a single item, otherwise referred to as a single point failure (SPF). A 2-order CS has two items ANDed together, a 3-order CS contains three items, and so on.

Cut Set Truncation

Truncation involves removing cut sets from the FT evaluation process to simplify and speed-up the computation. CSs are truncated when they exceed a specified order and/or probability. They are assumed to have minimal impact on the total FT probability.

Dependent Event/Failure

A dependent failure is when a conditional relationship exists between two components, whereby the failure of one is conditional upon failure of the other; the second failure depends on the first failure occurring. In probability theory, events are dependent when the outcome of one event directly affects or influences the outcome of a second event. To find the probability of two dependent events both occurring, multiply the probability of A and the probability of B after A occurs; P(A and B) = P(A) • P(B given A) = P(A) • P(B|A). This is known as conditional probability. Two failure events A and B are said to be dependent if P(A and B) ≠ P(A)P(B). In the presence of dependencies, often, but not always, P(A and B) > P(A)P(B). This increased probability of two (or more) events is why CCFs are of concern. It should be noted that in system design, a secondary failure is not necessarily a dependent failure, but more accurately a different root-cause for failure. For example, excessive heat on a transistor from an external source may cause the transistor to fail, but conditional probability is not used in this situation because it is not a dependency situation.

Event

An event is an occurrence that causes something to happen, such as a change of mode or state. In system safety and FTA an event typically refers to a failure or an anomaly. A component failure is an event whereby the component changes from the operational to the non-operational state.

Exposure Time (ET)

ET is the length of time a component is effectively exposed to failure during system operation. For example, all of the electronics equipment aboard an aircraft that is powered during a five-hour flight has a five-hour exposure time. ET has a large effect on FT probability calculations ($P=1.0 - e^{-\lambda T}$); as the exposure time increases the probability of failure increases exponentially. Exposure time can be controlled by design, repair, circumvention, testing or monitoring.

Failure

A failure is the inability of an item (e.g., system, subsystem, component or part) to perform its required functions within specified performance requirements. A failure may result in an unsafe condition (i.e. a hazard), the inability of a system element to perform a required function or it may cause a component to function incorrectly.

Fault

A fault is the occurrence, or existence, of an undesired state of an item; an undesired anomaly in the functional operation of a system, subsystem, component, item or part. Any change in state of an item that is considered to be anomalous may warrant some type of corrective action or safety protection. Faults are preliminary indications that a failure may have occurred. For example, "Light Off" is an undesired fault state that may be due to light bulb failure, loss of power or operator error. (Note – all failures are faults, but not all faults are failures). Distinguishing between fault and failure is an aid to the FT construction process.

Fault Duration Time (FDT)

This is the length of time a component remains in the failed state. This state is ended by repair of the component or by system failure. The FDT is dependent on the system design. If repair is possible during the mission, then the FDT will depend upon how long it takes for repair after the component fails. If repair is not possible, the FDT will likely be until mission completion. During this period of time the component can contribute to system failure or cause system failure.

Fault Tree (FT)

A FT is a logic model that diagrams the potential faults that can lead to the occurrence of a specified UE. It is a model of the cause-effect relationships leading to an undesired system event, containing all of the logic and events necessary and sufficient to cause the top UE to occur.

Fault Tree Analysis (FTA)

FTA is the analysis methodology that constructs a fault tree logic diagram of an UE, and then evaluates the causal factors and probabilities generated by the logic diagram. It utilizes tree structure methods, Boolean algebra and probability mathematics. Qualitatively it provides a list of cut sets that can cause the top UE to occur. Quantitatively it provides the probability of each cut set, the probability of the top UE and various importance measures for components and cut sets.

FT Pruning

FT pruning is the process of reducing the overall size of the FT in order to simplify the FT mathematical evaluation process. The FT is reduced in size by eliminating events in the FT, usually low probability non-relevant events. Pruning is generally achieved using FT truncation by CS order or CS probability.

Functional Block Diagram (FBD)

A FBD is a systems engineering tool for graphically showing how a system operates by linking system functions together. FBDs show system hierarchy, system interrelationships and system dependencies. FBDs are also known as Functional Flow Diagrams and Functional Dependency Diagrams.

Importance Measure (IM)

An IM is the measure of the relative importance of a component or CS in the overall FT. It establishes which failure events or CSs will influence the top FT probability the most with a change in the event failure rate or probability. IMs can show how sensitive the FT probability is to changes in the probability of an event or CS.

Independent Event/Failure

An independent failure is when the failure of a component is solely due to an inherent failure mode of the component, and is not caused by the failure of a different component or an external event. In probability theory, events are independent when the outcome of one event does not influence the outcome of a second event. System components are independent when the outcome of one component failure does not influence the outcome of a second component. To say that two events are independent means that the occurrence of one event makes it neither more nor less probable that the other component failure occurs. To find the probability of two independent events both occurring, multiply the probability of the first event by the probability of the second event; P(A and B) = P(A) • P(B). For example, the probability of tossing two dice and getting a 3 on each one is P(A and B) = P(3) • P(3) = (1/6) • (1/6) = 1/36, since the two events are independent. Most failure modes in a FTA are considered as independent failures.

Intermediate Event (IE)

An IE is any GE node in the FT structure, other than the top gate and the bottom level gates. IEs develop the cause-effect relationships that must be developed down the FT structure until the BEs are reached. An IE is also sometimes referred to as an Intermediate Gate Event (IGE).

Mission Time

The length of time the system is in operation to complete the mission. Most equipment is in operation during this period of time.

Multiple Occurring Branch (MOB)

A MOB is a FT branch (or sub-tree) that occurs in more than one place in a FT. For example, the failure of a power supply might be a FT branch that is relevant in several different areas of the FT. MOBs are indicated through the use of Transfer symbols. All events within the MOB branch are automatically MOEs, thus an MOB is a source of many MOEs. An MOB is a repeated branch that occurs in more than one place in the FT, and is therefore also known as a repeated branch.

Multiple Occurring Event (MOE)

A MOE is the same unique FT Basic Event that occurs in multiple places in the FT structure due to the particular system design and the FT structure corresponding to that design. It is not a similar event, it is a repeated event. For example, the event "Resistor 717 fails shorted" may naturally occur in the FT structure in several different locations. It is also known as a repeated event.

Non-coherent FT

A non-coherent FT is a FT that contains either a NOT gate or Exclusive OR (XOR) gate, since the XOR gate contains a hidden NOT. A non-coherent FT is significant in that more complicated mathematics are caused by the Not logic.

Primary Failure

A primary failure of a component is an independent failure that cannot be further defined at a lower level. It is an inherent failure mode of the component. For example, the failure modes "diode fails open" and "diode fails shorted" are typical inherent primary failure modes for diodes. The failure modes and failure rates of a component are generally determined by the physics, materials and/or manufacturing process of the component.

Prime Implicant

A prime implicant is a cut set for a non-coherent FT.

Probabilistic Risk Assessment (PRA)

A PRA is a quantitative evaluation that is performed to determine the probabilistic risk associated with an event, typically involving a complex system. The PRA provides the probability of an event occurring and the overall consequential severity of the event. Performing a PRA requires a systematic and comprehensive methodology, such as FTA. A PRA answers the questions of a) what are the possible undesired outcomes, b) what are the root causal factors involved and c) what is the risk probability presented.

Reliability

Reliability is the probability that a device will perform its intended function, without failure, during a specified period of time under stated conditions.

Reliability Block Diagram (RBD)

A RBD is a reliability model of a product or system that is used to understand and predict system reliability. It is a graphical and probabilistic model.

Repeated Branch

See Multiple Occurring Branch (MOB).

Repeated Event

See Multiple Occurring Event (MOE).

Secondary Failure

A secondary failure of a component is an externally induced failure that is caused by an external factor or force on the component, causing the component to fail by exceeding its design parameters. For example, "diode fails due to excessive RF/EMI energy in the system". In this example, excessive electromagnet energy on the component causes early failure of the component. A secondary failure is typically the result of out-of-tolerance operational or environmental conditions. A secondary failure is an independent component failure that is directly caused by a separate independent event as a root-cause. It is a cause-effect relationship; however, it is also a dependency relationship because the second failure would not have occurred if the first failure had not occurred. Some example causes of secondary failure include: temperature, RF energy, water, electrical voltage, etc. In most FTA quantitative calculations the dependency is ignored (they are treated as independent failures) because the final probability difference is small.

Sensitivity Measure

A sensitivity measure indicates how sensitive a component probability is to the overall FT probability. It helps to determine if a change in the component probability will have a significant influence the top FT probability. This can help establish whether a better failure rate is needed for a component in a quantitative analysis.

Single Point Failure (SPF)

A SPF is the failure of a single item or component that would, in turn, directly lead to the occurrence of a specified UE. Typically, the UE is a safety-critical condition, such as loss of life, loss of the system, loss of mission, environmental damage, etc. The particular UE depends upon the particular system and its objectives. It should be noted that from a very broad perspective a system could have hundreds or even thousands of SPFs in the design that could potentially cause anything from minor to catastrophic outcomes. The design concern is, which of the SPFs will cause a specified UE to adversely affect safety (or reliability) objectives. These SPFs present potential design safety weaknesses which should be eliminated or mitigated. A design statement that *all* SPFs will be eliminated is unreasonable; only those SPFs adversely impacting safety or reliability goals should be eliminated.

System

A system is a group of individual elements that interact and function together as a whole. It is an entity comprised of an integrated and interacting combination of elements to accomplish a defined objective. These elements include hardware, software, firmware, people, information, techniques, facilities, services, and other support items. The system as a whole is able to achieve more than any individual element of the system. A system may be viewed at various levels of complexity. A system is a collection of different things which, working together produce a result not achievable by the things alone; various elements working together for a larger purpose. Systems result in special qualities, properties, characteristics, functions, behaviors and performance unique to the system. Relationships and inter-connectivity of the parts is an important factor in system design and complexity.

System Safety

System safety is an engineering discipline for developing safe systems and products, where safety is intentionally designed into the system or product. It involves the planned application of management and engineering principles, criteria, and techniques for the purpose of developing a system that presents acceptable mishap risk. System safety applies to all phases of the system life cycle and covers all system aspects, such as hardware, firmware, software, human operators and procedures. System safety is the process for eliminating or reducing potential mishaps through a process of hazard identification, safety risk assessment and safety risk management. System safety is holistic and interdisciplinary in nature.

Top Undesired Event (TUE)

A TUE is the top node or event of the FT. It is a safety problem or hazard to be investigated and modeled. It is the starting point for the root-cause analysis that follows below it. The TUE defines and scopes the problem or hazard under investigation by the FTA.

Undesired Event (UE)

A UE is an event, or potential event, that is unwanted because of its undesirable safety consequence. Typically it is a safety problem or hazard that is being modeled by FTA to determine the potential root-causes of the postulated event.

CHAPTER 4

FTA PROCESS

4.1 The Basic FTA Process

FTA involves more than just constructing a FT diagram, it also requires many other tasks such as data collection, analysis, mathematical evaluation, etc. There are eight basic steps in the overall FTA process, as shown in Figure 4.1, that are required to perform a complete, thorough and accurate FTA. Some analysts may combine or expand some of the steps, but the overall process includes the following basic steps:

1) Define the system – Understand the system design and operation. Acquire current design data, such as: drawings, schematics, procedures, diagrams, etc.
2) Define the top UE – Descriptively define the problem and establish the correct UE for the analysis.
3) Establish FTA boundaries – Define the analysis ground rules and boundaries. Scope the problem and record all ground rules, boundaries, limitations, etc.
4) Construct the FT – Follow the construction process, rules and logic to build the FT model of the system for the specified UE.
5) Evaluate the FT – Generate the FT cut sets and probability from the FT and the input failure data. Identify weak links and safety problems in the design from the cut sets and probability.
6) Validate the FT – Check the FT model to ensure it is correct, complete and accurately reflects the system design.
7) Modify the FT – Modify the FT as found necessary during validation or due to system design changes that arise.
8) Document the FTA – Document the entire analysis with supporting data; provide the document as a customer product and preserve for future reference.

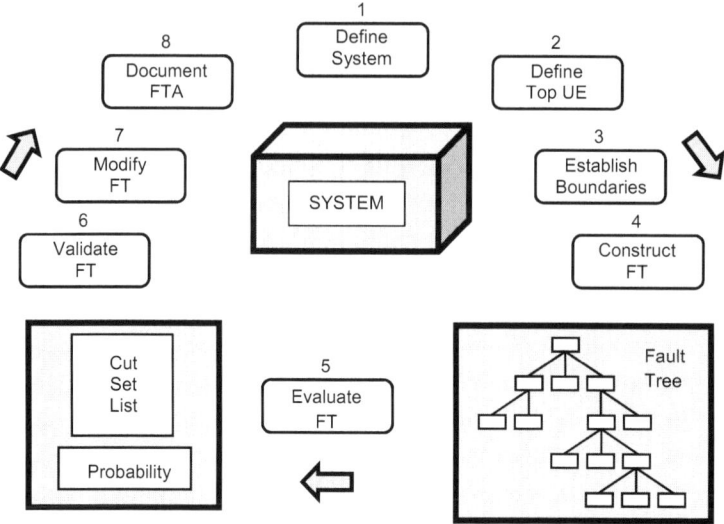

Figure 4.1 – Overall FTA Process

4.2 Define the System (Task 1)

To begin the FTA the system under investigation must be thoroughly defined and understood by the analyst. This involves acquiring current design data, such as: drawings, schematics, procedures, diagrams, etc. It also includes acquiring system development tools, such as: functional diagrams, reliability block diagrams, timelines, etc. In order to perform a credible FTA it is important to know and understand:

- System design
- System operation
- System components
- Software design and operation
- Hardware-software interactions
- Maintenance plans and operations
- Test plans and test procedures
- Manufacturing plans and processes

4.3 Define the Top Undesired Event (Task 2)

The next step is to determine the top undesired event (UE) for the analysis. UEs are generally established from hazard analysis or system design requirements. The top UE is typically an event of major safety concern; there is usually some unknown degree of mishap risk associated with it. The top UE may be a program-encompassing event, such as "mission loss" indicating an analysis that will further identify and develop specific potential mishap scenarios. The top UE may be specific, such as "crash due to engine failure" indicating an analysis for a specific potential mishap and subsystem. A specific hazardous condition, such as "inadvertent RF radiation occurs" may be the top UE, or it may be a specific hardware failure, such as "faults cause power supply failure". Defining the top UE for analysis is dependent upon the needs and objectives of the particular safety program.

The FT top UE is important because this is where the FTA effort begins. It shapes the entire analysis and must therefore be carefully selected and worded. An incorrect top UE may result in wasted analysis time and effort. Typically the UE is derived from a hazard analysis, design requirement, certification requirement, known safety problem, design concern or a mishap that has occurred.

Develop the top UE is using the following two steps: 1) write a long narrative description of the UE and 2) condense the long narrative to a short one-sentence description. The intent of the long narrative is to ensure the UE is fully understood, achievable and correct. It should be tested for validity and correctness. Program concurrence should be obtained on the final defined UE. After the long narrative UE is determined to be suitable, it should be condensed into a short statement that will fit in the FT text box.

It is important to make sure the UE is not too broad in scope such that the FTA would be too time consuming or impossible to achieve. For example, the UE of "all passengers aboard the tour ship die" is too broad in scope and should be narrowed down to achievable and safety-related potential events. Some example top UEs include the following:

- Failure of vehicle brake system to provide braking function
- Failure of the railroad vehicle collision avoidance system
- Inadvertent deployment of aircraft engine thrust reverser
- Loss of all aircraft communication systems
- Inadvertent weapon unlock
- Inadvertent weapon release

- Incorrect weapon status signals
- Offshore oil platform overturns during towing
- Failure of reactor emergency core cooling system

Top UEs should focus on potentially catastrophic events, where once the causal factors have occurred the system has reached a point of no return, and the UE will occur with certainty. These are events that the system design must prevent or reduce in probability of occurrence. For example, once faults have caused inadvertent bomb release or premature warhead detonation, there is no going back or no do-over.

4.4 Establish Boundaries (Task 3)

This step entails delineating the boundaries of the system, both internal and external, as well as boundaries for the analysis. A system baseline configuration must be established and controlled throughout the analysis, in order for the conclusions drawn from the analysis to be meaningful and valid. In addition, the limits or constraints to be placed upon the analysis must be defined. These limits may be in terms of resources, such as time or money, or they may be in terms of specific results desired. In addition, ground rules for the analysis should be established and documented. If the FTA ground rules are not documented, quite often they are forgotten halfway through the analysis. It is important to bound the overall scope of the problem to ensure the proper system elements and components are covered in order to ensure that the analysis includes the credible and critical failure modes and errors. Bounding the FTA works closely with step 2 of establishing the correct top UE. Boundary considerations include the following:

- System performance – areas of impact
- FTA approach – scenarios, functional paths, etc.
- Scope of analysis – what subsystems and components to include
- System modes of operation – startup, shutdown, steady state
- System phase or phases
- Available resources (i.e., time, dollars, people)
- Resolution limit (how deep to dig)
- Level of analysis detail and comprehensiveness

4.5 Construct the FT (Task 4)

Once the top UE is identified and the analysis ground rules are established, the FT is structured to model the system conditions that could cause the UE. This is done in accordance with the FT construction technique delineated in detail in Chapter 5. FT construction is an iterative process of continually checking the system design as the potential fault paths are followed through the system. It involves following the rules and definitions of FTA as the FT is developed in layers, levels and branches.

4.6 Evaluate the FT (Task 5)

When the FT structure is completed, an evaluation of the FT is performed to determine the results or significance of the analysis. Two types of evaluations are possible, qualitative and quantitative. The qualitative evaluation is an inspection and assessment of the FT cut sets (CSs). The quantitative evaluation is a numerical evaluation, where failure rates of the system elements are inserted into the FT structure and mathematically combined to yield probabilities. The purpose of the evaluation is to determine from the FT what risk is associated with the top UE and to identify which event, or events, are unacceptable and must be eliminated or controlled in order to eliminate or control the occurrence of the UE. Quite often in a program, the qualitative evaluation is conducted as the FT is being constructed to validate the FT accuracy, and then the quantitative evaluation is performed at the conclusion of the FT construction.

A qualitative analysis involves generating the CSs from the FT, verifying the correctness of the CSs and evaluating the CSs for safety implications. A CS consists of a single event, or combination of multiple events, that will cause the top UE to occur. Visual inspection of the CSs shows safety weak points in the system design. A quantitative analysis involves generating probability numbers and importance measures from the CSs. Probability calculations require the input of component failure rates and exposure times into the FT. By appropriately summing the probability of all the CSs a system level probability number can be derived. The system probability is an indication of the relative level of safety of the system in regard to the UE. Individual CS probabilities show which components and which fault paths are most critical in the system design configuration being analyzed.

FT evaluations can be performed manually or automatically via computer programs. Manual computations are really only feasible for small to medium size noncomplex FTs. Larger and more complex FTs require the use of a computer. FTA software solves FTs by direct analytical algorithms or via simulation. There are many different algorithms that have been developed for generating CSs and probability numbers, which are discussed further in Chapter 7.

Based upon the information derived from the FT evaluation, design decisions are made as to the adequacy of the system's level of safety. The design must be approved as safe, or the design problem areas identified and corrected. System safety corrective action is achieved by developing preventive measures which are recommended to design engineering in the form of design safety features. As recommended preventive measures are incorporated into the design, their adequacy in solving the safety problem must be verified. This is done by making the appropriate changes in the FT structure and then re-evaluating the FT. The effects of the change, or the relative measure of improvement, should be apparent from the FT re-evaluation.

4.7 Validate the FT (Task 6)

When the FT structure is completed and evaluation of the FT has been performed, all of the results should be inspected and analyzed to ensure the FT is correct and accurate in respect to system representation. That is, make sure the FT correctly models the system as designed and built. The following are various methods of validating the FT for correctness:

- Check each individual CS to make sure it will cause the UE
- Check the computed probability with manual calculations for reasonableness in the results
- Check all of the FT gate types to make sure one hasn't been accidently changed
- Review the failure rate data for accuracy
- Have subject matter experts peer review the FT
- Check all MOEs for correctness; ensure none were created by accidentally repeating a name
- Check to ensure there are no logic loops in the FTA
- Perform an intuition check; if something doesn't feel right check it out

Validation of the FT will ensure the quality of the results and will provide assurance to management and the customer the FTA was done correctly. Remember, computer programs do occasionally make mistakes; therefore all results should be double checked manually.

4.8 Modify the FT (Task 7)

It may be necessary to modify the FT if any errors were found during the FT validation (step 6). It may also be necessary to modify the FT if design changes were made to the system configuration. If changes are made to the FT make sure everything is documented appropriately so the history of the FT can be correctly recorded in case questions should arise at any point in time.

4.9 Document the FTA (Task 8)

One important aspect of FTA that is often overlooked is documentation. It is very important to document the entire FTA process for several reasons, such as:

- To provide credibility to the analysis
- To provide a formal customer product
- To provide a historical record for possible future use
- To retain the data in case a future update is necessary
- To provide an example FTA for a similar system that is under analysis

It is recommended that the FTA is documented using a formal documentation process established by the company performing the FTA. Documentation of a FTA should include the following elements as a minimum:

- FT structure
- FTA scope, boundaries and ground rules
- FT failure data with sources
- Design data used (drawings, drawing numbers, pictures)
- FT validation results
- FT evaluation results
- FTA conclusion and recommendations

CHAPTER 5

FTA CONSTRUCTION

5.1 Root-Cause Analysis

There is a universal truism that states: nothing happens without a cause. Root Cause Analysis (RCA) is any structured methodology for identifying the root-cause, or causes, of a problem. RCA is based upon following the cause-effect relationships involved from the problem start to the final *root* causal factor of the problem. RCA involves an iterative process of going from one set of cause-effect relationships to the next; the cause for one particular effect becomes the effect for the next causal factor. RCA involves asking "why" and "what-if" questions. Finding the root-cause is not typically achieved in the first set of questions, it typically requires asking a continuous set of questions, moving through a serious of cause-effect relationships, until pay dirt is struck.

RCA is typically a reactive process of identifying event causes after an event (or mishap) has occurred. However, when done pragmatically RCA can be used as a proactive process to forecast or predict probable causes even before they occur. Logic tree type diagrams are effective methodologies in RCA. In this respect, FTA is a logic tree analysis and diagramming tool that very effectively identifies and links cause-effect relationships until the root-causes of an UE are discovered. Figure 5.1 shows how the causal factor at one level in the FT become the effect for the next level. When complete, one can trace through the FT structure and see the logical cause-effect relationships linked together into a single coherent FT. The FTA construction process provides the logic, tools and structured methodology for asking all of the right questions needed to identify the cause-effect relationships in a system design that ultimately lead from the root-causes to the UE. The specific what-if questions to ask in FTA are presented in the following sections.

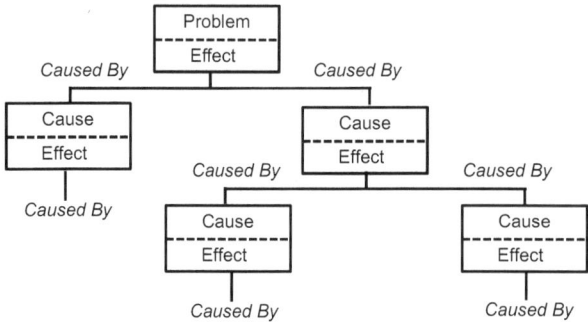

Figure 5.1 – Cause and Effect Relationships

5.2 FTA Is a Cause-Effect Process

FT construction is an iterative process that begins at the FT top UE and continues down through all of the tree branches. The same set of questions and logic is applied at every FT gate, moving down the tree. After identifying the top UE, sub-undesired events are identified and structured into what is referred to as the *top fault tree layer*. The actual deductive analysis begins with the development of the *fault flow* or cause and effect relationship of fault and normal events through the system. This deductive reasoning process involves determining the type of gate and the particular inputs to this gate at each gate level of the FT. The fault flow links the flow of events from the system level, through the subsystem level, to the component level. The FT development proceeds through the identification and combination of the system normal and fault events, until all events are defined in terms of basic identifiable hardware faults, software faults and human error. The end of the FT is reached when the basic root-causes can no longer be broken down to a lower level.

FTA is a root-cause analysis methodology that seeks to determine all the possible root-causes leading to an UE. The FT structure becomes similar to a linked chain, where the causal events are linked to the effect events in a chain-like manner, and the FTA is searching for the weak links in the chain. In the FT, a cause at one level becomes an effect for the next level of analysis, as shown in Figure 5.2, thereby linking the fault flow through the system.

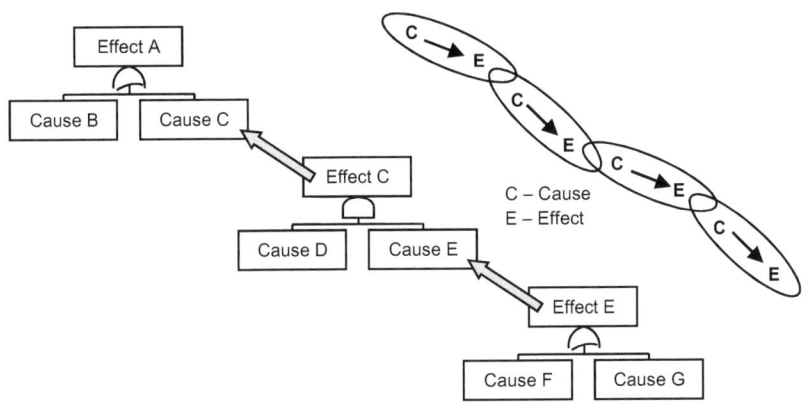

Figure 5.2 – FT Cause and Effect Chain

The idea in FTA is to proceed logically through the system and identify all of the root-causes than can lead to the top UE. This process involves establishing logic gates and the inputs to these logic gates. FTA involves many *cause-effect* relationships. Each FT gate is essentially an effect, for which the causes must be determined. Figure 5.3 depicts the concept, that for each gate the *how* and *what* questions must be answered. What are the causal events for event X and how are they logically combined together? The answer to these questions establishes the input events and gate type.

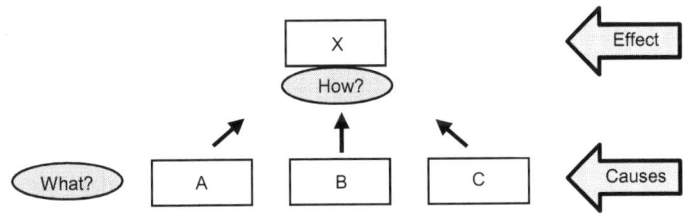

Figure 5.3 – FT Cause and Effect Determination

5.3 FT Structural Approaches

There are four basic approaches to developing a FT structure: Functional, Shotgun, Subsystem, and Scenario. The Functional approach logically follows the system functions. This approach develops the overall command path through the system. It has more levels, is more structured and is less prone to miss events. The Shotgun approach is when the analyst looks at the system and starts linking fault events together without extensive consideration to logical links between fault states and causal events. The Subsystem approach emphasizes subsystems (sometimes prematurely). It tends to have fewer levels and is broader across. The Scenario approach emphasizes scenarios (sometimes prematurely). It too tends to have fewer less levels and is also broader across. The Functional approach is the most effective and desired method, while the Shotgun approach is the least effective, especially for large complex systems, and the least desired methodology.

Figure 5.4 roughly depicts the Functional approach concept. The idea is that the analysis begins at component E, which is involved in the beginning of the UE, and follows the path backwards against the functional flow of the system.

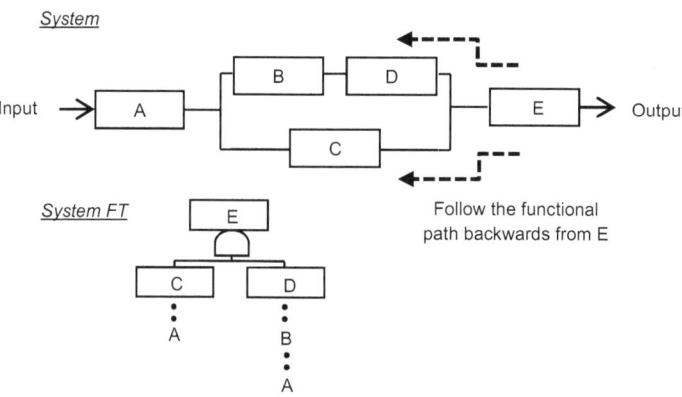

Figure 5.4 – FTA Functional Flow Approach

Figure 5.5 depicts the Shotgun approach concept. The idea is, that as system components that could cause the top UE are identified by visual inspection, they are immediately placed in the FT at the highest level possible. This approach uses very little analytical examination of the system fault logic that is involved. Note the difference between this FT and the one in Figure 5.4. The Shotgun approach tends to miss and overlook faults events when the system is large and complex; as such it is not recommended. Some individuals use it because it can be done quickly and it avoids having to apply logic skills. It is also referred to as the Shopping List approach.

Fault Tree Analysis Primer

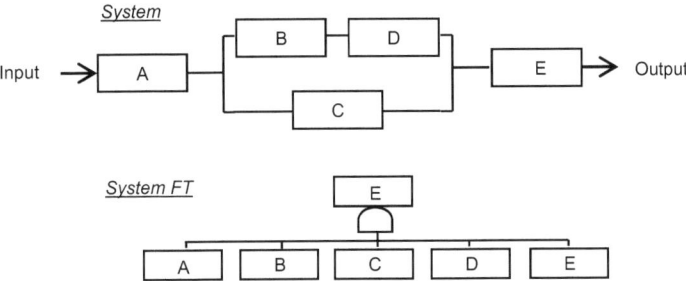

Figure 5.5 – FTA Shotgun Approach

Figure 5.6 depicts the Subsystem approach concept. The idea is that the FT is broken-down, at the top level by subsystems. Each subsystem's unique contribution to the top UE is then determined. This approach is sometimes used when subsystem probability allocations are needed.

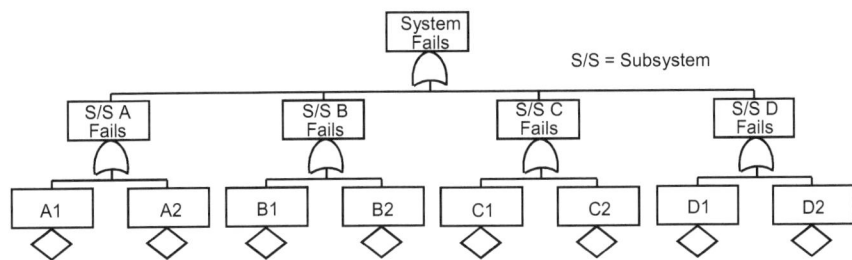

Figure 5.6 – FTA Subsystem Approach

Figure 5.7 depicts the Scenario approach concept. The idea is that as system components that could cause the top UE are identified, they are immediately placed in the FT at the highest level possible.

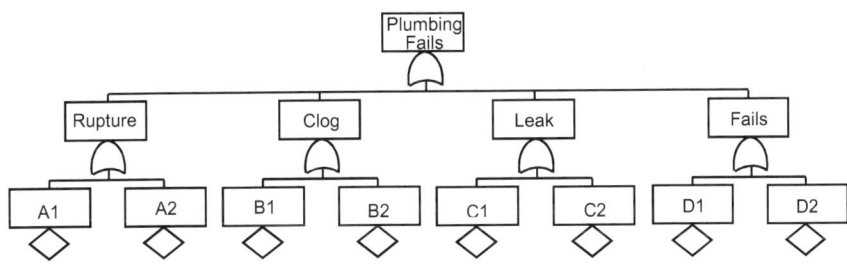

Figure 5.7 – FTA Scenario Approach

5.4 FT Construction Methodology

In developing the structure of the FT, a special thought process exists that is very productive in consistently identifying gate types and gate input events. It is an analytical process applied in a repetitive manner. This process involves asking three sets of questions, at each FT gate, which examine the system from three different perspectives. These perspectives and questions are referred to as[1]:

1) The I-N-S View
2) The SoS-SoC View
3) The P-S-C View

When the information obtained from these sets of questions is combined, the gate logic type will have been determined and the input causal factors will have been identified.

5.4.1 The I-N-S View

This concept involves answering the question "What is Immediate, Necessary and Sufficient (I-N-S) to cause the event?" The I-N-S question identifies the most immediate cause(s) of the event; the causes that are absolutely necessary; and only includes the causes that are absolutely necessary and sufficient. For example, water is necessary to maintain a green lawn and rain is sufficient to provide it, or a sprinkler system is sufficient to provide the water.

This seems like an obvious question to ask, but too often it is forgotten in the turmoil of analysis. There are several reasons for stressing this question:

1) It helps keep the analyst from jumping ahead
2) It helps focus on identifying the next element in the cause-effect chain
3) It's a reminder to include only the minimum sufficient causes necessary and nothing extraneous

5.4.2 The SoS-SoC View

The SoS-SoC concept differentiates between the failure being a "State-of-the-System" (SoS) or a "State-of-the-Component" (SoC) type failure. If a fault in the gate text box can be caused by a component failure, then

---

[1] NUREG-0492, Fault Tree Handbook, N. H. Roberts, W. E. Vesely, D. F. Haasl & F. F. Goldberg, 1981, 208 pages, U.S. Government Printing Office.

classify the event as a SoC fault. If the fault cannot be caused by a component failure, classify the fault as a SoS fault. If the fault event is classified as SoC, then the event will have an OR gate with multiple inputs, which are determined from P-S-C logic, discussed below. If the fault event is classified as SoS, then the event will be further developed using I-N-S logic to determine the inputs and gate type.

5.4.3 The P-S-C View

This concept involves answering the question "What are the Primary, Secondary, and Command (P-S-C) causes of the event?" The P-S-C question forces the analyst to focus on specific causal factors. The rationale behind this question is that every component fault event has three ways of failing: a primary failure mode, a secondary failure mode and a command path fault mode. Figure 5.8 demonstrates this concept of how a system element is sub-divided into primary, secondary, and command causal factors for the FT structure. The command fault establishes a gate event that must be further analyzed. A benefit of this concept is that if more than two of the three elements of P-S-C are present, then an OR gate is automatically indicated.

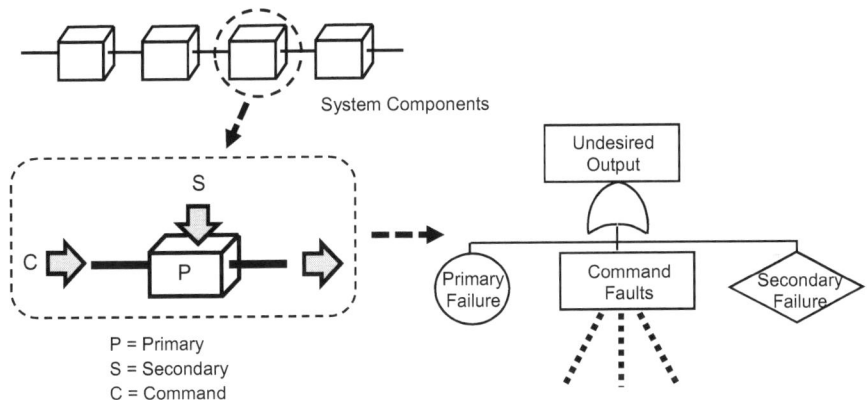

Figure 5.8 – P-S-C Concept

A primary failure is the inherent failure of a system element (e.g., a resistor fails open). The primary failure is developed only to the point where identifiable internal component failures will directly cause the fault event. The failure of one component is presumed to be unrelated to the failure of any other component (i.e., independent).

A secondary failure is the result of external forces on the component (e.g., a resistor fails open due to excessive external heat exposure).

Development of the secondary failure event requires a thorough knowledge of all external influences affecting system components (e.g., excessive heat, vibration, EMI, etc.). The failure of one component may be related to the failure of other components (i.e., dependent). This type of component failure is due to any cause other than its own primary failure.

A command failure is an expected, or intended, event that occurs at an undesired time due to specific failures. For example, missile launch is an intended event at a certain point in the mission. However, this event can be *commanded* to occur prematurely by certain failures in the missile arm and fire functions. Failures and faults in this chain of events are referred to as *command path* faults.

The command path is a chain of events delineating the path of command failure events through the system. Analysis of command path events creates an orderly and logical manner of fault identification at each level of the FT. A path of command events through the FT corresponds to the signal flow through the system. In developing command events, the question "what downstream event commands the event to occur" is asked for each event being analyzed. At the finish of each FT branch, the command path will terminate in primary and secondary events.

Note that the command path is primarily a guideline for analysis of fault event development through a system. Once an analysis is completed, comparison between the FT and the system signal flow diagram will show that the FT command path represents the signal flow through the system along a single thread.

For another example of a command path fault, consider a relay. When the relay coil is energized, the relay contacts will automatically close, as designed and intended. If a failure downstream of the relay provides inadvertent power to the relay coil, then the closing of the relay contacts is considered as a "command" failure. The relay operates as normally intended, except at the wrong time.

5.5 FT Construction Iteration

As previously mentioned, FT construction is a repetitive process. Figure 5.9 displays this iterative process, where for every logic gate on the FT, the same set of three questions is asked: I-N-S, SoS/SoC and P-S-C. Answering these questions provides the gate input events and the gate logic involved. As seen from this diagram, the iterative analysis proceeds downward, while the cause-effect relationships are linked in an upward manner.

Fault Tree Analysis Primer

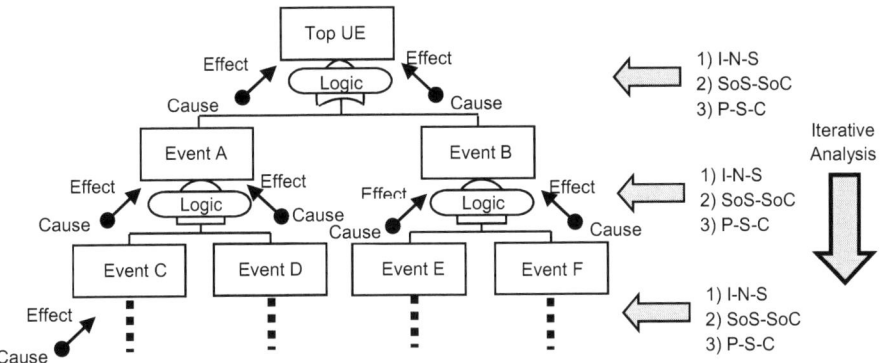

Figure 5.9 – FT Building Steps

The basic steps to follow when constructing the FT include the following iterative process:

1) Review and understand the fault event (i.e., FT gate) under investigation
2) Identify all the possible causes of this event via the questions:
   a) Is it *Immediate, Necessary and Sufficient?*
   b) Is it *State of System or State of Component?*
   c) What are the *Primary, Secondary and Command causes?*
3) Identify the relationship or logic of the cause-effect events
4) Structure the FT with the identified gate input events and gate logic
5) Double check logic to ensure that a jump in logic has not occurred
6) Keep looking back to ensure identified events are not repeated
7) Repeat for next fault event (i.e., gate)

Some critical items to remember while performing the FT construction process include the following:

1) When possible, start analyzing in the design at the point where the UE occurs
2) Work backward (through system) along the functional signal path or logic flow
3) Keep node wording clear and precise, with sufficient detail to understand the logical thought process

4) Check to ensure all text boxes have unique text; no repeated text should occur
5) Ensure you do not jump ahead of a possible fault event in the system structure
6) Look for component or fault event *transition* states (e.g., "no output signal from component A", "no input fluid to Valve V1")

Figure 5.10 is a diagram summarizing the overall FT construction process. It serves as a reminder chart of things to consider when performing a FTA.

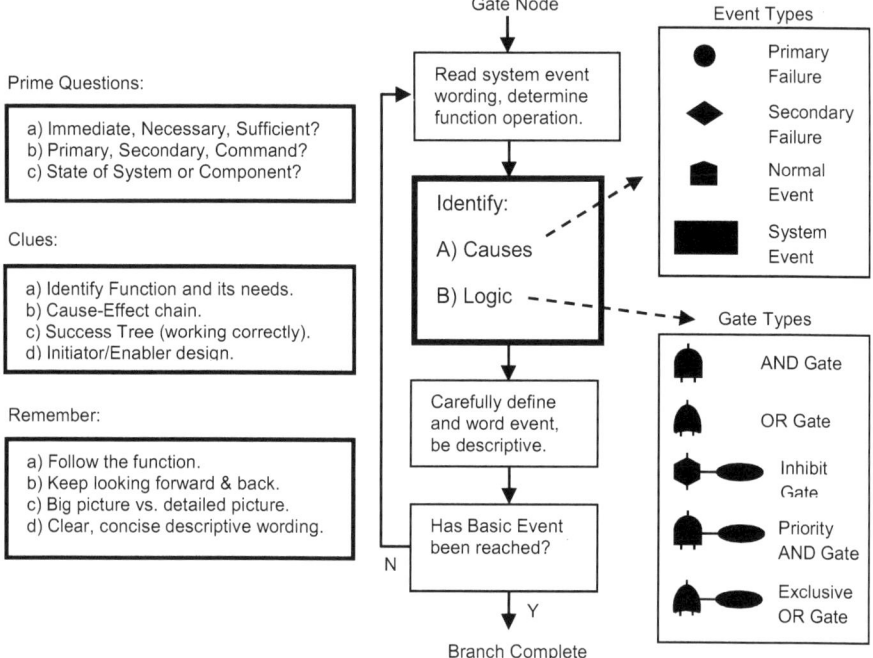

Figure 5.10 – Summary of Overall FT Process

CHAPTER 6

FTA CONSTRUCTION EXAMPLES

6.1 Simple FTA Examples

In order to demonstrate FT logic and FT construction, two simple examples are provided. The first example is a lighting system with two switches in series and the second example is a similar lighting system with the two switches in parallel.

6.1.1 Series System

Figure 6.1 shows the system with two switches in series and the corresponding FT. In this example it is assumed both switches are closed by the operator to turn the light on. The UE in this case is "light is off". The FT model shows the system failure modes that would cause this situation. In this FT any component failure will cause the UE, therefore it uses all OR gate logic.

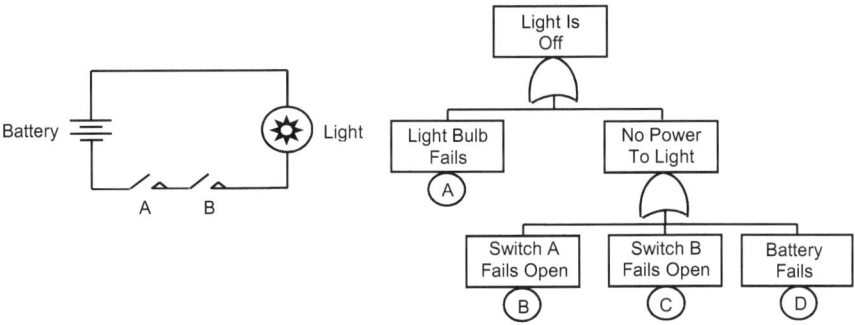

Figure 6.1 – FT Model of Switches in Series

6.1.2 Parallel System

Figure 6.2 shows the system with the two switches in parallel and the corresponding FT. In this example it is assumed both switches are closed by the operator to turn the light on. The UE in this case is "light is off". The FT model shows the system failure modes that would cause this situation. This FT requires both OR and AND gate logic. The two switches are in parallel, or are redundant, as known in system design terminology. In order to defeat a redundant design both components must be failed at the same time, thereby necessitating an AND gate.

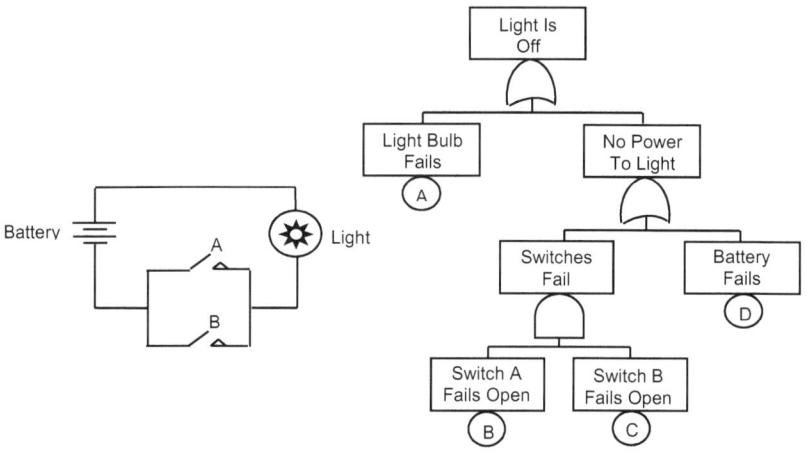

Figure 6.2 – FT Model of Switches in Parallel

6.2 Intermediate FTA Examples

The following are some FTA examples provided to demonstrate FT logic and FT construction for intermediate level system complexity. The first example is the FT of a monitor system, the second example is the FT of a Standby system and the third example is an aircraft fire suppression system.

6.2.1 Monitor System

Figure 6.3 shows the FT of a monitor system. This system is comprised of two components, Monitor A and component B. Monitor A monitors the operation of B, however, it is only designed to monitor 80% of B. If the monitor detects any failure in B it takes corrective action. System success requires that B must operate successfully. System failure occurs if component B fails, which can only happen if Monitor A fails to detect a problem with the monitored portion of B, or if the unmonitored portion of B fails. The darker portion of component B represents the portion of B that is not monitored ($\lambda B1$). If B fails before A the system is designed to take corrective action, therefore A must fail first, otherwise there will be no system failure. Note that this is a summary FT and many of the events would be expanded during the actual analysis.

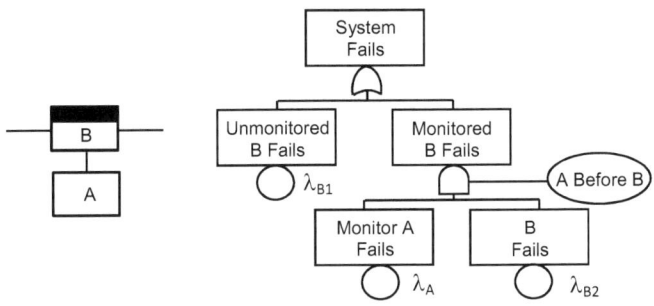

Figure 6.3 – Monitor System FT

6.2.2 Standby System

Figure 6.4 shows the FT of a standby system. This system is comprised of two redundant components A and B, and a monitor. System operation starts with component A in operation and B in standby mode. If A fails, then B is switched on-line and it takes over. System success requires that either A or B operate successfully. System failure occurs if both components A and B fail. Note that B can be failed if switching fails to occur due to various faults. There are three classes of Standby systems: Hot Standby where backup is powered during standby (uses operational λ_O), Warm Standby where standby is partially powered during standby ($\lambda_W < \lambda_O$) and Cold Standby where backup is un-powered during standby ($\lambda_C = 0$). Note that this is a summary FT and many of the events would be expanded upon during the analysis.

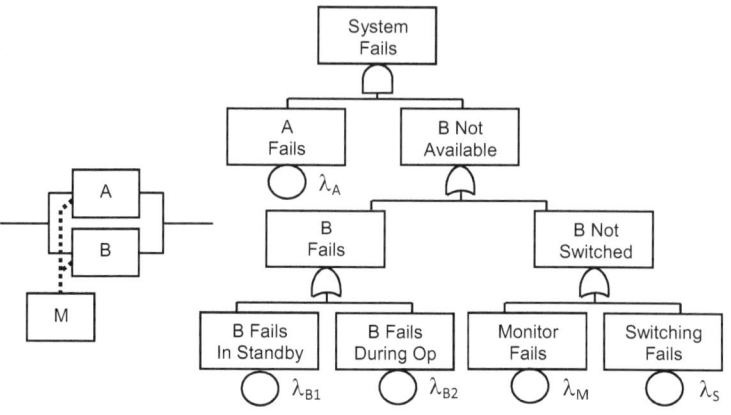

Figure 6.4 – Standby System FT

6.2.3 Fire Detection System

Figure 6.5 shows a FTA of a system designed to detect and suppress a fire aboard an aircraft. The FT events with a diamond symbol can be further expanded in a more detailed analysis. This FT demonstrates how the fire occurs, combined with failure of the fire detection or fire suppression system.

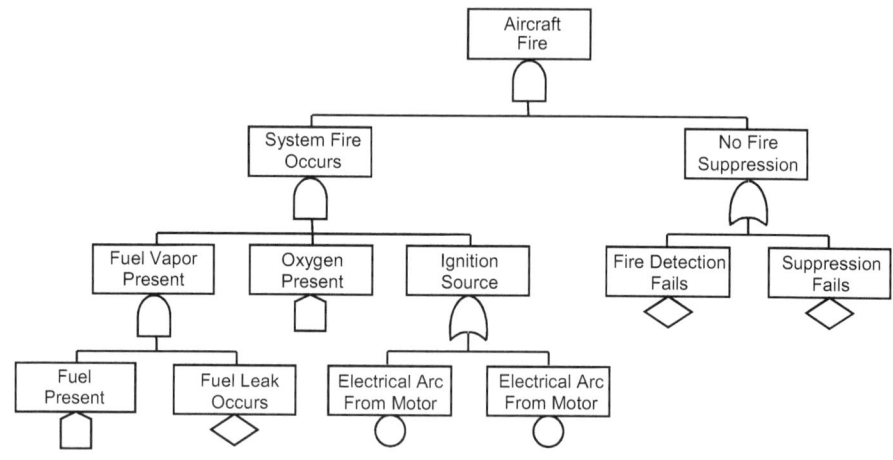

Figure 6.5 – Fire Detection System FT

Note in this example FT how the House symbol indicates an event that is normally expected to occur, as a result of the system design, during system operation. Typically the probability of this type event is 1.0, but it could be adjusted to something less than 1.0 when necessary.

6.3 Complex FTA Examples

The following are some examples provided to demonstrate FT logic and FT construction methods for complex situations. The first example is the FT for an UE of fire on an aircraft. The second example deals with the UE of probability of loss of aircraft (PLOA). These examples are more complex because they analyze a large system, where it is unclear how to start the top levels of the FT.

6.3.1 Aircraft Fire

Figure 6.6 shows the FT for an UE of fire on an aircraft. This FTA takes a large problem and breaks it down into manageable pieces. This is the top level of the FTA and each branch would be analyzed deeper into the subsystems and components. At this high level in the FT structure there may be alternative approaches to takes, before getting down into the hardware level.

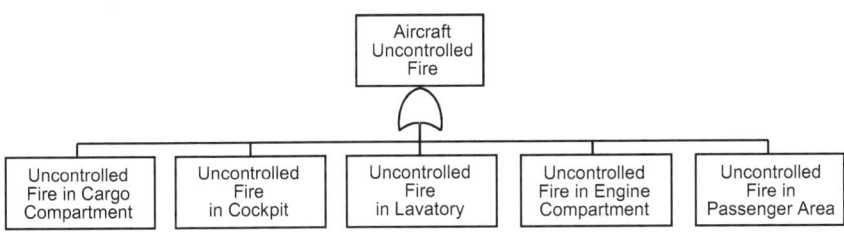

Figure 6.6 – Aircraft Fire FT

6.3.2 Probability of Loss of Aircraft (PLOA)

Figure 6.7 shows the FT for an UE of probability of loss of aircraft (PLOA). This is a type of FTA performed on manned and unmanned aircraft. One purpose for this FTA is to verify that aircraft subsystems meet reliability requirements. Another use for this FTA is to determine if the likelihood of an aircraft loss is within acceptable ranges and to identify the weak links in the system. This is the top level of the FTA and each branch would be analyzed deeper into the subsystems and components. At this high level in the FT structure there may be alternative approaches to takes, before getting down into the hardware level. Note that CFIT stands for "controlled flight into terrain".

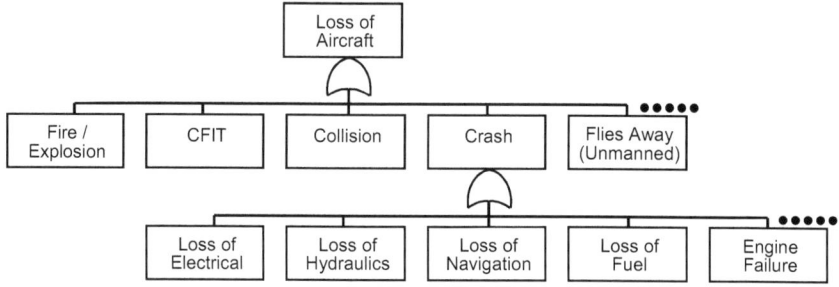

Figure 6.7 – Probability of Loss of Aircraft (PLOA) FT

In this PLOA example, the dotted lines indicate that there may possibly be more events to add to the FT.

CHAPTER 7

FTA MATHEMATICS

7.1 Introduction to FTA Mathematics

FTA mathematics encompass the mathematical methods and algorithms used to qualitatively and quantitatively evaluate a FT. Basically, there are three aspects of FT evaluation: 1) FT cut set generation, 2) FT probability generation and 3) FT importance measure generation. The qualitative and quantitative aspects of FTA are based primarily on the following mathematics, which form the foundational building blocks for FTA evaluation:

- Probability --- establishes FT gate logic and FT probability
- Boolean Algebra --- resolves non-minimal and extra FT CSs
- Reliability --- establishes FT component probability of failure

FTs can be qualitatively and quantitatively evaluated, either manually or via a computer. Manual evaluation is acceptable for small FTs, but for large and complex FTs a computer and a FT computer program are really needed, since manual FT evaluation is very tedious and error prone. One nice feature of FTA is that the same mathematics applies to both manual and computer evaluation. Figure 7.1 shows a very generic gate-to-gate evaluation process for deriving a FT probability using the FT mathematical methods, each of which are described in more detail. In this diagram the gate probability calculation is a function of the gate type, which determines the specific probability calculation to use.

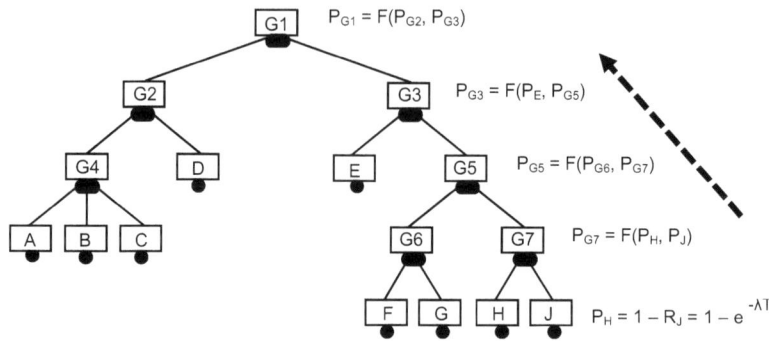

Figure 7.1 – FT Gate-to-Gate Probability Example

7.2 Probability

One of the main mathematical components of FT evaluation involves probability theory, which is used to calculate the probability for each FT gate, including the top gate. One method to calculate the top FT probability is by starting at the bottom of the FT and moving towards the top, performing probability calculations for each gate (gate-to-gate method). The particular formula used for each gate is a function of the gate type. However, when MOEs and MOBs are in the FT this methodology no longer works correctly, because the MOEs must be resolved to their lowest minimum set; which is typically achieved using Boolean algebra reduction.

Another method to calculate the FT top probability is to first generate all of the minimum CSs and then summing the probability for each CS using the appropriate probability equation.

Table 7.1 lists the basic laws or rules of probability. These probability rules translate into the probability formulas for the various FT gate types, which are defined in the following sections.

Fault Tree Analysis Primer

Table 7.1 – Laws of Probability

	Rule Statement
R1	The probability of an event is between 0 and 1; $0 <= P(E) <= 1$
R2	If an event is certain to occur, then $P(E) = 100\% = 1.0$ If an event is certain not to occur, then $P(E) = 0\% = 0$
R3	It is certain that an event will either occur or not occur, therefore $P(E + \text{not } E) = P(E) + P(\text{not } E) = 1$ or $P(\text{not } E) = 1 - P(E)$

R4	The additive property of two probabilities is:
	a) If two events are disjoint (i.e., both cannot happen)
	   $P(E1 \text{ OR } E2) = P(E1) + P(E2)$
	
	b) If two events are non-disjoint (i.e., they can occur together)
	   $P(E1 \text{ OR } E2) = P(E1) + P(E2) - P(E1 \bullet E2)$

R5	The multiplicative property of two probabilities is:
	a) if two events are mutually independent
	   $P(E1 \text{ AND } E2) = P(E1) \bullet P(E2)$
	
	b) if two events are not mutually independent (interdependent)
	   $P(E1 \text{ AND } E2) = P(E1) \bullet P(E2|E1) = P(E2) \bullet P(E1|E2)$

Figures 7.2 and 7.3 show the probability formula, Venn diagram and truth table for the AND gate and the OR gate, respectively.

$P = P_A \bullet P_B = P_A P_B$

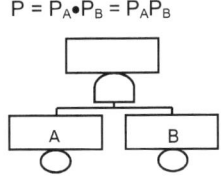 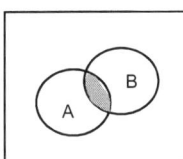

A	B	TOP
0	0	0
1	0	0
0	1	0
1	1	1

Figure 7.2 – AND Gate Probability Formula

$P = P_A + P_B - P_A P_B$

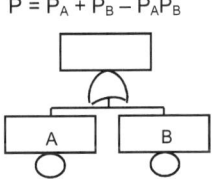 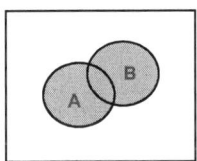

A	B	TOP
0	0	0
1	0	1
0	1	1
1	1	1

Figure 7.3 – OR Gate Probability Formula

The gate probability formulas shown for the AND and OR gates, in Figures 7.2 and 7.3 respectively, are for two input gates. When there are more than two inputs to these gates the basic formulas must be expanded accordingly. The AND gate expands very easily, whereas the OR gate expansion formula is a little more complicated. For example, look at the following equations for a two, three and four input gate:

Two input gate:
 OR: $P = P_A + P_B - P_{AB}$
 AND: $P = P_A \times P_B$

Three input gate:
 OR: $P = P_A + P_B + P_C - (P_{AB} + P_{AC} + P_{BC}) + P_{ABC}$
 AND: $P = P_A \times P_B \times P_C$

Four input gate:
 OR: $P = P_A + P_B + P_C + P_D - (P_{AB} + P_{AC} + P_{AD} + P_{BC} + P_{BD} + P_{CD}) + (P_{ABC} + P_{ABD} + P_{ACD} + P_{BCD}) - (P_{ABCD})$
 AND: $P = P_A \times P_B \times P_C \times P_D$

The reason the OR gate expansion is more complicated can be seen in the Venn diagram shown in Figure 7.3. For a two input OR gate the intersection area is added into the equation twice, therefore one intersection area must be subtracted out. For a three input OR gate the intersection area is added into the equation three times, thus some must be removed. However, this time too much is removed, therefore a smaller area must be added back. This process continues in a similar manner for each additional input to the OR gate.

7.3 Boolean Algebra

Boolean algebra and probability theory are used together to generate a list of CSs from the FT failure events and gates events. A typical FT can produce many CSs, depending on the FT size and complexity. MOEs, MOBs and certain AND-OR gate combinations produce non-minimal and duplicate CSs. In order to generate the final correct list of CSs produced by the FT structure, the total list of CS combinations must be reduced to minimal CSs, or MinCSs as they are typically called. Boolean reduction of the FT is achieved by applying the laws of Boolean algebra.

Table 7.2 provides the Axioms (A) and Theorems (T) of Boolean algebra. The theorems and axioms with an X-mark next to them are the ones that are most used in FT CS reduction.

Table 7.2 – Boolean Algebra Axioms and Theorems

	Axiom/Theorem	FTA
A1	ab = ba	X
A2	a + b = b + a	X
A3	(a + b) + c = a + (b + c) = a + b + c	
A4	(ab)c = a(bc) = abc	
A5	a(b+c) = ab + ac	
T1	a + 0 = a	
T2	a + 1 = 1	
T3	a • 0 = 0	
T4	a • 1 = a	
T5	a • a = a	X
T6	a + a = a	X
T7	a • not a = 0	
T8	a + not a = 1	
T9	a + ab = a	X
T10	a(a + b) = a	X
T11	a + (not a)(b) = a + b	

7.4 Reliability

Reliability mathematics is used to calculate the probability of failure for the basic FT input events (component failures). Once calculated, the FT input failure event probabilities are applied to the appropriate formula to produce a gate or CS probability. The basic reliability equations that are used in FTA are:

- $R = e^{-\lambda T}$
- $R + Q = 1$
- $Q = 1 - R = 1 - e^{-\lambda T}$
- Approximation: $Q \approx \lambda T$ when $\lambda T < 0.001$

Where:
  R = Reliability or probability of success
  Q = Unreliability = P or probability of failure
  λ = component failure rate = 1 / MTBF
  T = time interval (mission time or exposure time)

Time has a big impact on the probability of failure of a component. The longer the component exposure time the higher the resulting probability of failure. Exposure time refers to the length of time the component is operational or powered. As the exposure time increases, the probability of failure increases exponentially, approaching P = 1.0. Conversely, a shorter exposure results in a smaller probability of failure. The component failure rate also effects probability, the smaller the failure rate the lower the probability of failure and the larger the failure rate the higher the probability of failure. Figure 7.4 demonstrates the effect of time on the probability of failure. In this example, a component is given a constant failure rate of 1.0 x 10^{-6} failure per hour (FPH) and the exposure time is varied from 1 hour to 10 million hours. The increasing probability of failure is very noticeable.

λ (FPH)	Time (Hrs)	$P_A = 1 - e^{-\lambda T}$
1.0xE-6	1	9.99xE-7
1.0xE-6	10	9.99xE-6
1.0xE-6	100	9.99xE-5
1.0xE-6	1,000	9.99xE-4
1.0xE-6	10,000	9.95xE-3
1.0xE-6	100,000	0.095
1.0xE-6	1,000,000	0.6321
1.0xE-6	10,000,000	0.99995

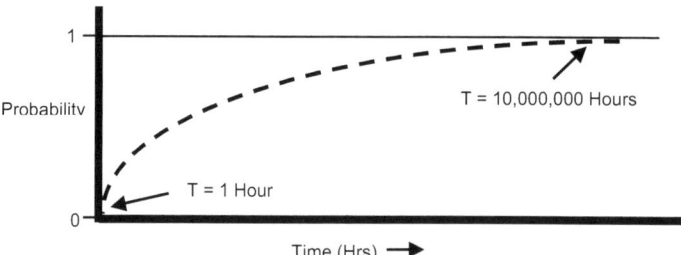

Figure 7.4 – Effect of Time on Event Probability

Component exposure time can be either an advantage or disadvantage since it tends to drive the probability of failure for a component. When the FT top probability is too high, reduce the probability of failure of critical components by reducing their exposure time through a system design change.

7.5 Gate Probability Formulas

Table 7.3 summarizes the probability formulas for the basic FT gates. These are the formulas that define the gate function.

Table 7.3 – Gate Probability Formulas

Fault Tree	Venn Diagram	Set Equation and Probability Equation
AND gate with G, A, B	AND	$G = A \times B$ $P(G) = P(A) \times P(B)$
OR gate with G, A, B	OR	$G = A + B$ $P(G) = P(A) + P(B) - P(A)P(B)$ Non-disjoint case
Priority AND gate with G, A, B	Priority AND	$G = A \times B$, With A before B $P(G) \approx 1/n! \times P(A) \times P(B)$ Given $\lambda_A \approx \lambda_B$
Exclusive OR gate with G, A, B	Exclusive OR	$G = A + B$, but not both together $P(G) = P(A) + P(B) - 2(P(A)P(B))$ Mutually exclusive
Inhibit Gate with G, A, B	Inhibit Gate	G is same as AND P is same as AND

7.6 Cut Sets

Cut set (CS) is a term used in FTA that identifies a unique set of events that, together, can cause the top UE of the FT to occur. The elements or events comprising a CS can be failures, human errors, software anomalies, environment conditions or normal system actions. A CS can have any number of events in it; for example a CS could be comprised of one event or 14 events. Multiple events within a CS indicate that the events are ANDed together. A large FT can have well over 300,000 cut sets or more.

A *minimal cut set* (MCS), sometimes referred to as MinCS, is a cut set where none of the set elements can be removed from the set and still cause the top event to occur. A minimal cut set is a combination of events just sufficient for the top UE to occur. The combination is a minimal combination in that all the failures are needed for the top event to occur; if one of the failures in the cut set does not occur, then the top event will not occur (by this combination of events). All FTs will consist of a finite number of minimal cut sets that are unique for that top event and FT.

Unique systems designs can result in FTs that produce duplicate CSs and non-minimal CSs. In order to calculate a correct FT top probability, all CSs must be reduced to minimal form. Figure 7.5 shows an example FT and the CSs derived from it; note that a 1-order CS is a single point failure. In this example, all CSs are minimal without applying Boolean reduction.

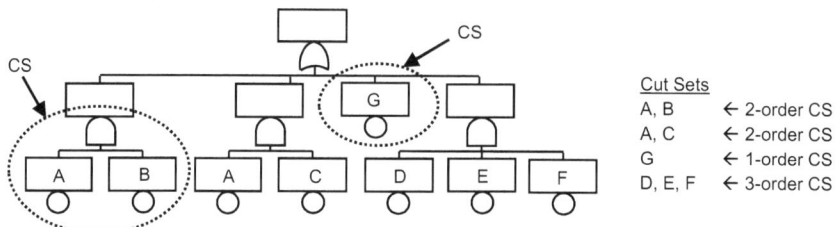

Figure 7.5 – FT with Cut Sets Shown

CSs are one of the most valuable products from FTA. CSs identify the root-causes of the top UE. They also identify critical components and critical system failure paths. One method of obtaining the top FT probability is by summing all of the CSs, using the appropriate equation. CSs are an aid to decision making; they help identify the weak links in the system design and they help identify places in the system design where safety improvements can best be made.

CS *order* refers to the number of elements within a cut set. Usually, the higher the cut set order the lower the CS probability, since CSs of order 2 or greater represent AND gate conditions where the input probabilities are multiplied together. CS *truncation* refers to truncating or eliminating certain CSs from the overall FT probability calculation in order to make the calculation less complex and faster. Truncation is typically done by affecting the CS probability or order. For example, all CSs with a probability of 10^{-9} or smaller, or of order 6 or greater, are dropped from the overall calculation (because their contribution to the top probability would most likely be negligible). Truncation parameters should be carefully selected and the final calculation results carefully checked, as truncation can sometimes introduce errors or misunderstandings.

There are several different methods available for determining CSs, such as:

- Boolean reduction; manual or computerized
- Top down reduction algorithms, such as MOCUS (Method of Obtaining Cut Sets)
- Bottom up reduction algorithms, such as MICSUP (Minimal Cut Sets Upward)
- Binary Decision Diagram (BDD) algorithm
- Modularization methods
- Min Terms method (Shannon decomposition)

In the process of generating CSs for a FT, quite often many unneeded CSs are generated. This is because unique systems designs can result in FTs that produce duplicate CSs and non-minimal CSs. When this happens CS resolution is required to reduce the total set of CSs to only the set of MinCSs. Figure 7.6 shows some example FTs and why certain branches are eliminated due to the laws of Boolean algebra. The FT program will automatically resolve the CSs; the FT does not have to be redrawn to show events eliminated from CSs.

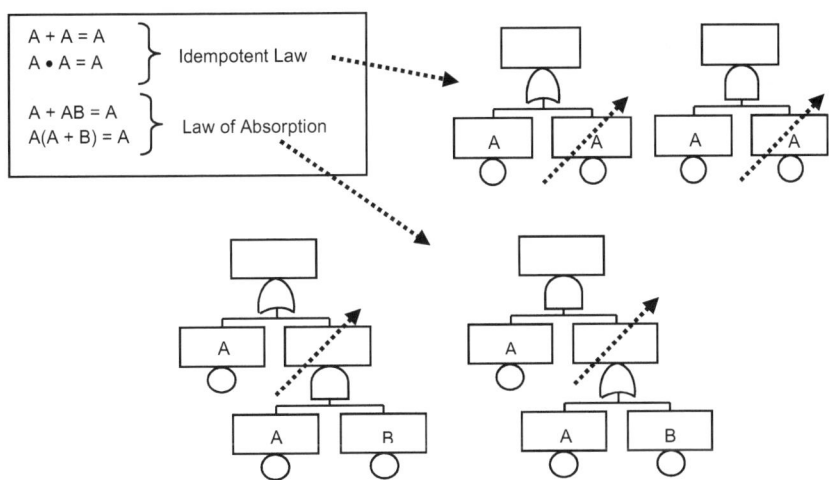

Figure 7.6 – FT Reduction by Boolean Algebra Laws

7.7 Cut Set Reduction

One of the main purposes of representing a FT in terms of Boolean equations is that these equations can then be used to determine, or resolve, the minimal CSs in the FT sets. Once the minimal CSs are obtained, the quantification of the FT is more or less straightforward.

The minimal CS expression for the FT top can be written in the general form of:

$$T = M_1 + M_2 + ... + M_k$$

where T is the top FT event and M1, M2, M3, etc. are the minimal CSs. Each minimal CS consists of a combination of specific component failures, and hence the general n-component minimal CS can be expressed as:

$$M_i = X_1 \bullet X_2 \bullet ... \bullet X_n$$

where X_1, X_2, etc., are basic component failure events within the CS. The minimal CSs are unique for each FT.

To determine the minimal cut sets of a FT, the FT is first translated to its equivalent Boolean equations. A variety of algorithms exist to translate the Boolean equations into cut sets. Two of the most common are the top-down or bottom-up substitution methods to solve for the top event. The methods are straightforward and involve substituting and expanding Boolean expressions.

The following are some example CSs produced from theoretical FTs. Note that the "+" sign separates the individual CSs and the "," sign denotes a CS where all of the elements are ANDed together and the comma separates the elements within the CS. The final MinCSs in each case are derived by applying the Boolean algebra laws.

1) A + B + A,C + A,D = A + B
2) A,B + A,C + E + E,A,C + F = A,B + A,C + E + F
3) X + R + X,Y + X,R + X,Y,R = X + R

Figure 7.7 shows an example of top-down Boolean reduction for an example FT to obtain the minimal CSs.

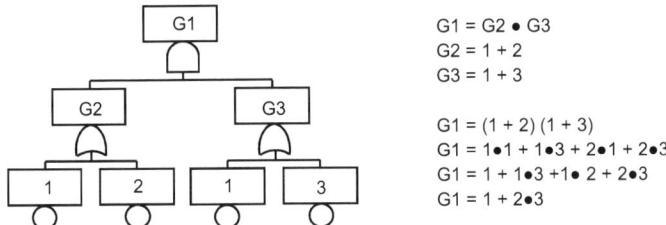

Figure 7.7 – Boolean Reduction Example

This Boolean reduction example demonstrates the overall methodology of the process. This example was fairly simple for three failure events with one MOE. Think how long and difficult the process would become for a FT of 1,000 failure events with many MOEs.

7.8 MOCUS CS Algorithm

One of the most common FT algorithms for generating CSs is the MOCUS (Method of Obtaining Cut Sets) algorithm, developed by J. Fussell and W. Vesely[2]. This is an effective top-down gate substitution methodology for generating CSs. MOCUS is based on the observation that AND gate increase number of elements in a CS and that OR gate increase the number of CSs.

The basic steps in the MOCUS algorithm are as follows:

1) Name or number all gates and events.
2) Place the uppermost gate name in the first row of a matrix.
3) Replace top gate with its inputs, using the following rules:
   a) Replace an AND gate with its inputs, each input separated by a comma.
   b) Replace an OR gate by vertical arrangement, creating a new line for each input in the OR gate.
4) Reiteratively substitute and replace each gate with its inputs, moving down the FT.
5) When only basic inputs remain in the matrix, the substitution process is complete and the list of all CSs has been established.
6) Remove all non-minimal CSs and duplicate CSs from the list using the laws of Boolean algebra.
7) The final list contains the MinCSs.

---

[2] J. B. Fussell, W. E. Vesely, A New Method for Obtaining Cut Sets for Fault Trees, 1972, Transactions ANS, No. 15, p262-263.

Figure 7.8 provides an example of applying the MOCUS algorithm to a FT. This example FT consists of three gates and three input events, where one of the input events is an MOE. The MOE requires Boolean reduction. This example FT was resolved in six steps; however, larger FTs would naturally require more steps.

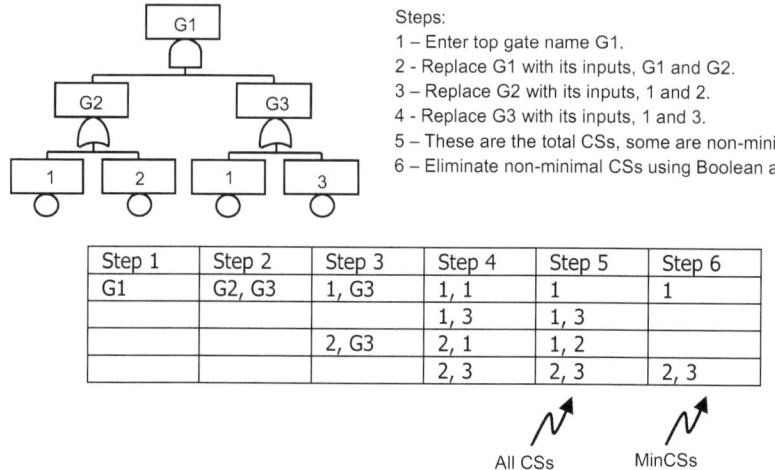

Figure 7.8 – MOCUS Example

7.9 Quantitative Approximation Algorithms

7.9.1 OR Gate Expansion Formula

One method of calculating the top FT probability is by *summing* all of the MinCSs together using the OR gate formula. Since all of the CSs are individual root-causes, they can essentially all feed into one big OR gate. The expansion formula for an OR gate is shown below. This expansion formula is shown for four CSs only (A, B, C and D), but it would be expanded accordingly for more CSs. Each CS in this formula represents a single CS. For example A is a single CS, which could be comprised of one or more elements, such as E1•E2 (i.e., E1 AND E2).

$$P = P_A + P_B + P_C + P_D - (P_{AB} + P_{AC} + P_{AD} + P_{BC} + P_{BD} + P_{CD}) + (P_{ABC} + P_{ABD} + P_{ACD} + P_{BCD}) - (P_{ABCD}) \bullet \bullet \bullet$$

7.9.2 Inclusion-Exclusion Approximation Method

A FT can have any number of CSs, depending upon the FT size and complexity. In order to obtain the correct FT probability, all of the MinCSs must be ORed together using the expansion formula shown in Figure 7.9.

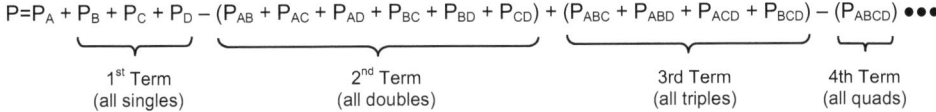

Figure 7.9 – Inclusion-Exclusion Approximation

If a FT had 100,000 CSs, the expansion formula would to be applied to 100,000 elements. This would become a very computer intensive calculation to perform. In order to simplify FT software programming and reduce computer time, two approximation algorithms have been developed to streamline the process. The first approximation technique is called the Inclusion-Exclusion approximation method. This method breaks the OR gate expansion formula into sections, called 1st term, 2nd term, 3rd term, etc. as shown in Figure 7.10. It has been recognized that the 1st term is actually the upper bound for the exact probability and the 2nd term is the lower bound. The more terms that are applied in the calculation, the closer the result come to the exact probability. Since FTA typically involves very small numbers, calculating to the 2nd or 3rd term usually provides close enough accuracy.

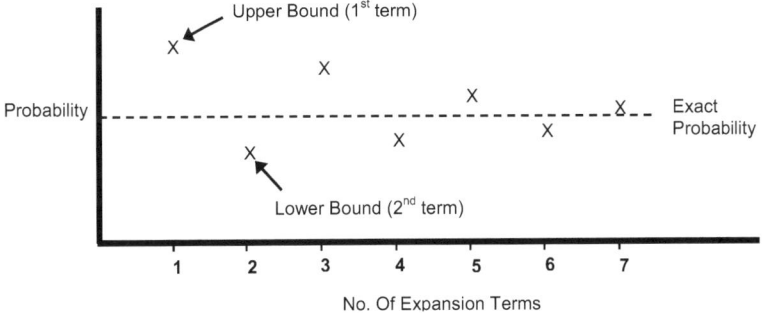

Figure 7.10 – Inclusion-Exclusion Bounds

7.9.3 Min Cut Set Upper Bound Approximation Method

The second approximation technique is called the Min Cut Set Upper Bound approximation method. In this method all of the MinCSs are determined, and then the probability of each CS is calculated. The top FT probability is then computed using the following formula:

$$P = 1 - [(1 - P_{CS1})(1 - P_{CS2})(1 - P_{CS3})\ldots(1 - P_{CSN})]$$

This method computes a very close approximation to the exact top FT probability, and in some cases, depending on the FT, it produces the exact solution. For example, for a FT with two CSs, A and B, the formula proves to be exact, as shown in the following calculation:

$$\begin{aligned} P &= 1 - [(1 - P_A)(1 - P_B)] \\ &= 1 - [1 - P_B - P_A + P_A P_B] \\ &= 1 - 1 + P_B + P_A - P_A P_B \\ &= P_A + P_B - P_A P_B \quad \text{(note – two input OR gate formula)} \end{aligned}$$

7.9.4 Comparison of Methods

The accuracy of these approximation methods is demonstrated in Figure 7.11, where the probability of an example FT is calculated using the true formula and the two approximation methods. This FT has four basic inputs, one of which is an MOE. The MinCSs are derived using the MOCUS algorithm and the FT probability is then calculated using the event probability data provided. Each basic event in this example is given a probability of $P = 0.1$. It can be seen from this demonstration how close the results from the approximation methods are to the actual result.

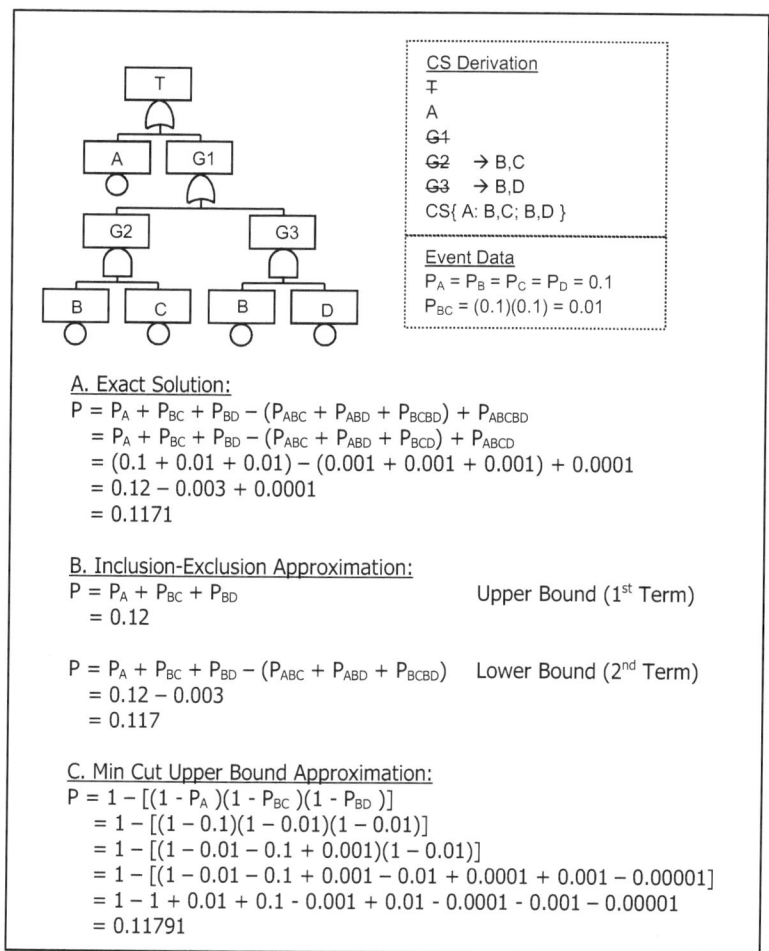

Figure 7.11 – FT Probability Comparisons

7.10 Importance Measures and Sensitivity Measures

Importance measures are an essential component of FTA. Importance measures establish the relative significance of all the events in the FT in terms of their contributions to the top UE probability. CSs, as well as basic fault events, can be prioritized according to their importance, which helps identify weak links in the system design and prioritize where design action is best applied.

The CS importance measure indicates the contribution of a specific CS to the total FT probability. When all the CS importance measures are ranked together, the significance of each CS becomes visible. An example of this calculation is as follows, where P_2 represents the probability of component 2 and P_T represents the probability of the total FT:

Min CS List
2
2, 3
2, 4 $\}$ ⟶ $I_{2,4} = (P_2 \times P_4) / P_T$
7, 8, 9, 10
4, 5, 6

The Fussel-Vesely importance measure indicates a specific component's contribution to failure in relation to the total FT probability. When all components are ranked together, this measure provides an analysis of the most critical components. An example of this calculation is as follows:

Min CS List
2
2, 3 $\}$ ⟶ $I_2 = ((P_2) \times (P_2 \times P_3) \times (P_2 \times P_4)) / P_T$
2, 4
7, 8, 9, 10
4, 5, 6

Sensitivity measures are also an essential element of FTA. They determine the sensitivity of the top FT probability to an increase or decrease in the probability of any event in the FT. Sensitivity analysis is performed to determine the effect of changing the failure rate or probability of a particular component. Sensitivity analysis shows which system components are most influential to the top FT probability and will therefore have the largest impact in changing the top FT probability.

The *Risk Reduction Worth (RRW)* sensitivity measure evaluates the decrease in the top FT probability if a given event is assured not to occur. The RRW value is calculated by re-quantifying the FT with the probability of the given event set to 0 (zero), while the other event probabilities are kept constant. The RRW indicates whether reducing the events probability will have an impact on the overall probability. This shows whether an improvement in the components failure rate will be beneficial or not. If the components failure rate is questionable to begin with, it also shows whether or not obtaining a better failure rate is worth the effort.

The *Risk Achievement Worth (RAW)* sensitivity measure evaluates the increase in the top FT probability if a given event occurs 100% of the time. The RAW value is calculated by re-quantifying the FT with the probability of the given event set to 1.0, while the other event probabilities are kept constant. Components failures with largest RAW value will have the largest system impact, therefore, these are the failures that should be prevented. The RAW shows where design mitigation activities should be focused in order to assure failures don't occur.

7.10 Important FTA Equations to Remember

The following equations are provided as a reminder for use during FT evaluation calculations:

- Event A probability:
  - $P_F(A) = P_A = 1 - e^{-\lambda T} \approx \lambda T$
  - $P_S(A) = R(A) = R_A = e^{-\lambda T}$
- AND gate probability (2 inputs):
  - $P = P_A \bullet P_B = P_A P_B$
- OR gate probability (2 inputs):
  - $P = P_A + P_B - P_A P_B$
- Boolean functions:
  - $a + a = a$
  - $a \bullet a = a$
  - $a + ab = a$
  - $a(a + b) = a$

CHAPTER 8

FTA RULES AND GUIDELINES

8.1 Rules and Guidelines

Of the many FTs I have reviewed over the years I have seen good FTs, mediocre FTs and bad FTs. Always strive to construct a good FT, otherwise the results may be worthless. Understanding FT rules and intentionally designing a FT helps to create a quality FT product. Don't just quickly produce a mediocre FT in order to meet a deadline or contract data item. A good FT shows credibility; it looks professional, thorough, complete and understandable. It's easy to visually inspect the quality of a FT; an experienced FT analyst can tell how much effort and credibility was put into a FT just through a visual inspection.

There are some basic rules for FT construction, which if followed, will help in the development of high quality FTs. These rules are broken down into the following categories:

- General rules
- Construction rules
- Evaluation rules
- Data rules
- Human error rules
- FT Design Aid Rules

8.2 General Rules

FTA requires analysis, logic and thoughtful consideration of the system. The following general rules provide an aid in the overall FTA process:

a) Know the system under investigation well; it is imperative to know and understand the system design and operation thoroughly. Utilize all sources of design information, such as: drawings, procedures, block diagrams, flow diagrams, FMEAs, stress analyses, failure reports, maintenance procedures, system interface documents and concept of operation. Drawings and data must be current for accurate results.

b) When constructing a FT, it is often much easier to work from a functional block diagram (FBD) than from a large complex electrical schematic to help shape the overall FT logic. After the basic logic is shaped, the electrical schematic is used for the detailed analysis. A general FTA rule of thumb states: if an analyst cannot draw a FBD of the system being analyzed, then the analyst may not fully understand the system design and operation well enough to perform a proper FTA.

c) Understand the purpose of the FTA; it's important to know why the FTA is being performed. This will help to ensure adequate resources are applied, a proper scope of analysis and that the appropriate results are obtained. Remember, FTA is a tool for root-cause analysis and probability analysis. It is also used to measure the relative impact of a design fix.

d) Establish a FTA standard for the project prior to beginning the FTA. This standard should provide consistent guidelines for the FTA project. Some example items to consider include guidelines for:

- The inclusion or exclusion of human error, software, CCFs, etc.
- The depth of analysis (subsystem, LRU, component level)
- The use of a naming convention for FT events and gates
- The approach for a component database (including failure rates and data sources)
- The approach for handling MOEs and MOBs

e) Understand the potential size of the FT, because FT size impacts the entire FTA process. As FTs grow in size many factors are affected, such as: cost (e.g., manpower), analysis time, complexity, understanding, traceability and computation. System factors that cause FT growth include: system size, system complexity and the safety criticality of system. FT factors that cause FT growth include: MOEs and MOBs (e.g., redundancy) and certain AND / OR gate combinations.

f) Intentionally design the FT; foresight and planning helps in avoiding future problems. As a FT grows in size it is important to develop a FT architecture and a set of FT development rules. The architecture lays out the overall FT design, which can aid in the analysis. For example, certain FT branches may apply to specific subsystems and specific subcontractors. On very large FTs certain branches can be assigned to different analysts.

g) FTA projects require planning, organization, structure and foresight, due to tree size, complexity and FT team communications. Proper FT design or architecture can help prevent major rework for small changes and errors in the FT structure or data. Considerations include: allowing for future expansions in the FT, allowing for ease of modification and providing for accounting and control of event and gate names. Strict control of MOE/MOB names is particularly import for correct results.

h) Ensure the FT is correct and complete; if the FT is not correct and complete the results will not be accurate. Anything left out of the FTA skews the answer. The final result will only reflect what was included in the FT. The FTA is not complete until *all* root-causes have been identified. Conduct a FT peer review to ensure completeness and correctness; involve other FT experts and system designers. Items often overlooked in FTA are human error, common cause failures, software factors, design dependencies and components or subsystems considered not applicable to the analysis.

i) Know and understand the computerized FT tool being used because tools, and tool misuse, can easily cause errors and they can provide a source of confusion. Understand the basic capabilities of the FT tool, such as construction, editing, plotting, report generation and cut set evaluation. Select a FT tool that is user friendly, provides intuitive operation and easily allows for FT changes. Know and work within the limitations of the FT tool. Limitations include tree size (i.e., max number of events), cut set size and print size. Understand the FT tool approximation and cutoff methods because some FT tools can cause errors if the algorithms are incorrect. Don't place complete trust in a FT tool; test the tool, don't assume answers are always correct.

j) Study the FTA results, fully understand them and make sure they are correct. Considerations include the following:

- Was the analysis objective achieved?
- Are the results meaningful?
- Was FTA the right tool?
- Are adjustments to the FTA necessary?
- Are the results correct?
- Are CSs credible and relevant (if not revise tree)?

Take nothing for granted from the computer results. Test the results via manual calculations; make reasonableness tests to verify the results. Look for analysis errors (logic, data, model, computer results).

k) Document the entire FTA process, including the FT model, assumptions, design data utilized, failure data and data sources, computation results and analysis conclusions. Even a small FTA should be documented for posterity. Reasons to document the FTA process include:

- It might be necessary to provide the customer with a formal FTA product
- It might be necessary to defend the FTA results at a later date
- It might be necessary to modify the FTA at a later date for design changes
- It might be used to support a similar system analysis at a later date
- It might be used to support an accident/incident investigation
- It could support future questions or analyses

8.3 Construction Rules

FT construction requires analysis, logic and thoughtful consideration of the system. The following rules provide assistance in the construction of good FTs:

a) Think in Terms of *failure space*; remember, it's a *fault* tree, not a *success* tree. The purpose is to evaluate failures, faults, errors, sneak paths and unsafe designs. Some analysts have difficulty switching their mode of thought from success space to failure space and they attempt to include successful operation in the FT.

b) Do not draw the FT assuming the system can be saved by a miraculous failure or an operator save. This is normally referred to as the "No Magic Rule". No operator *saves*; when constructing FT logic do not assume that operator action will save the system from fault conditions, only built-in design safety features can be considered. Operator errors can be considered in the FT, but not operator saves. The system design is under investigation, not the operator performing miracles.

c) Correct wording in the text boxes is important in order to convey the FT logic at each event and gate. The wording should be clear, precise and descriptive. Fault events should be expressed in terms of device transition or as an input or output state. For example, rather than saying "Power supply fails" it is much more descriptive and informative to say "Power supply does not provide +5 VDC". Or, "Valve fails in closed position" rather than "Valve fails". It should be noted that the terms Primary, Secondary and Command are a thought process and should not be used in the event description; the FT symbols used will automatically convey this information.

d) Apply FT aesthetics as much as possible. When the FT structure looks good it will be more easily understood and therefore more likely to be accepted. For example, a level and balanced FT structure looks best; do not use zigzags in the structure. When making printed page breaks avoid too little information on a page (i.e., 2 or 3 events only). Always use standard FT symbols (defined in NUREG book; reference 1). Computerized FT construction tools provide better graphics than manual methods.

e) Construct the FT to most accurately reflect the system design and logic. Do not try to modify the tree structure to resolve an MOE. Let the FT computer software handle all MOE resolutions.

f) When the words in a gate box exceed the box limit, it is allowable to create another input with a gate box directly below just to continue the words. This creates a single input OR gate, which is okay, but prudence is also necessary as single input OR gates should be kept to a minimum. Single input OR gates do not look good and they add to the computation complexity. FT notes can also be used if additional words are needed.

g) Use House events carefully. A House (Normal event) never goes into an OR gate, except in special cases, such as turning branches on and off. Turning branches on/off is a way of producing one FT for many different modes of system operation. This may work, but it is complex and can hide the FT results that are needed for the verification process.

h) Do not label fault events on the FT as *Primary*, *Secondary* and *Command* failures. These terms are used as an analysis thought process. The terms I-N-S, P-S-C and SoS-SoC are analysis thought processes; do not use these words in text boxes. The FT symbols will denote primary, secondary and command fault events.

i) When writing text in the FT text boxes, be as detailed and descriptive as possible. When practicable add traceability detail; put drawing numbers and part numbers in the fault event or in the notes. This provides better traceability when the FT is being reviewed or checked, or when the FT is being modified after a lengthy time period.

j) Operator error should be included in the analysis where appropriate. Remember, anything omitted from the FT skews the final results. It is up to the analyst and the objectives of the FTA as to whether certain events should be included or excluded from the analysis. This decision needs to be documented in the FTA ground rules.

k) Always take a second look at all FT structure and logic. Sometimes what appears to be a simple and correct FT logic and structure might actually be flawed for various reasons. For example, mutually exclusive events and logic loops can easily occur in the FT structure. Make sure there are no leaps or gaps in logic. The FT structure may need revising in these cases; this is normal in FT construction, do not expect perfection the first time.

l) Every event and gate in the FT must contain specific necessary information, which includes:

- Node name
- Node type
- Event failure rate (or probability)
- Event exposure time
- Node text (never leave a text box blank, otherwise the FT is meaningless)

m) When writing the words for a text box, state the event fault states exactly and precisely; use state transition wording when possible. For example, if water is not going into a valve, then "water is not input to Valve V1" is preferred to "Valve V1 fails".

n) No gate-to-gate FT diagrams are allowed. These types of diagrams omit the text box for one or more gates. Without the information in the text box the FT cannot be validated. See figure 8.1 for a visual explanation.

o) Do not draw a gate without a gate node box and associated descriptive text and rectangle. See figure 8.1 for a visual explanation.

p) Use only one output from a node. Do not connect the output of a node to more than one input nodes. Some analysts attempt to show redundancy this way, but it becomes cluttered and confusing. Most computer codes cannot handle this situation anyway. Use MOEs. See figure 8.1 for a visual explanation.

Figure 8.1 demonstrates some typical violations of the FT construction rules. Violation of these rules creates many problems. For example, if a text box is missing or has no text in it, then anyone reading the FT will be unable to understand or follow the logic involved. Do not draw lines from two different gates to a single input, use the MOEs and transfer symbols.

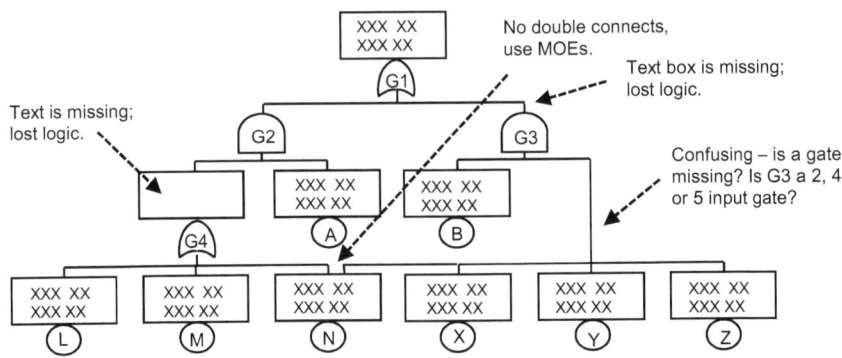

Figure 8.1 – Example FT Construction Errors

8.4 Evaluation Rules

FT evaluation consists of calculating the FT CSs, probabilities and importance measures. The following guidelines provide assistance in the FT evaluation process:

a) Computerized evaluation is preferred over manual calculations. Manual calculations are tedious and error prone, especially for larger and more complex FTs.

b) Check all CSs manually to ensure they are valid. Invalid CSs can skew the results significantly. This means comparing the CS to the system design and ensuring the CS will cause the UE.

c) Perform a numerical reality check on all probability results. Manually make some rough calculations to ensure the final results are in the right ball park.

d) Verify that all MOEs and MOBs are correct. They can have a significant impact on the final results. It is very easy to accidentally create MOEs and MOBs.

e) Remember, FTs are only models. The results are close approximations from a model and should not be taken as the exact answer.

f) Don't get carried away with numbers; understand and appreciate small numbers. Numbers in the range of 10^{-6} and smaller can tend to be difficult to visualize and comprehend. Keep a proper perspective on numbers.

g) A FTA is often used as a probabilistic risk assessment (PRA). Be sure to balance PRA results with risk reality. Determine if the PRA results are realistic when compared with similar hazards, similar systems and historical mishap rates.

h) Remember, FT accuracy depends on inclusion. Anything left out of the FT model will directly impact the accuracy of the final results.

8.5 Data Rules

FT quantitative evaluation requires failure rate data in order to calculate FT probabilities. The following guidelines provide an aid in FT data collection and data usage:

a) Understand the failure data; make sure it is reasonable and reliable.

b) Even raw data is valuable when reliable failure data cannot located. Sometimes it is useful to use raw, worst case data; if the resulting probability calculations are acceptable, then it may not be necessary to refine the data.

c) Sensitivity analysis can help determine if a more refined failure rate is really needed for a particular component.

d) Whenever using failure data, always provide the data sources. This adds credibility and provides a mechanism for tracking the data at some point in the future.

e) It is very helpful to use a database that contains all of the events and their associated failure data. All component failure rates should be included, along with their sources.

8.6 Human Error Rules

Human error can be a significant factor in system hazards and mishaps. Ignoring human error in the FTA could have undesirable consequences and skew the final results. The following guidelines provide an aid in using human error in a FTA:

a) The human is a key element in system operation, and therefore in system design. Human error should be included in the FT model when appropriate. The effects and criticality of human error on system operation can be determined in the same manner as is done with hardware.

b) Failure rate data for human error exists, but is often not applied because the rates are too high and they produce the dominant CSs in a FTA. The overall goal of the analysis will dictate whether or not the human element is included in the FTA, and whether or not human errors are quantified. When possible, quantify human error in the FTA. Remember, the real goal is to produce a safe system.

c) Decisions regarding the inclusion or exclusion of human error in the FT should be documented in the FTA ground rules. The same applies to human error failure data.

d) Do not include human action in the FTA to prevent an accident; no magical saves are allowed. The goal is to identify design weaknesses and flaws.

e) Human reliability and human error is a complex issue and Human System Integration (HSI) is a design safety concern. When including human reliability in the FTA, items to consider include:
- Operator fails to perform function
- Operator performs function incorrectly
- Operator performs function inadvertently
- Operator performs wrong function
- Operation action exacerbates the results of a system failure

8.7 FT Design Aid Rules

FT construction and evaluation can be assisted through the use of certain FT design aids. The following FT design aids have been proven to aid in the FTA process:

a) The use of short event and gate names can actually simplify the FTA process. Short names are much easier to write and track than long names. Using five character names along with special naming conventions provides the means to name 10,000 gates and events on the FT. Using a 24-character name can become confusing and tends to clutter the FT diagram.

b) Apply some forethought when naming FT events. The use of event naming conventions can be very beneficial. Specific characters are used to quickly identify event types. For example, the first character in the name can be used to identify the type of event:

- G = gate
- X = circle or primary failure
- Z = diamond or secondary failure
- W = house or normal event
- Y = oval or condition

The second character can be used to establish a pattern for tree branch families, for example:

- A = Computer System
- B = Guidance and Navigation System

Characters 3, 4 and 5 can be used to numerically identify the node, for example:

- XA1 through XA999
- GA1 through GA999

c) Event accounting is important in FTA, especially for very large FTs. It is very helpful to use a database that contains all of the events and their associated data. Failure rates should be included, along with their source. A FTA database can help in keeping track of FT node names and their location in the FT.

d) FT roadmaps are sometimes necessary for very large FTs. A FT roadmap is a small diagram that condenses the large FT, which is usually on many separate sheets of paper. The condensed FT does not contain all of the relevant FTA information, but it shows the overall FT structure. In this manner, the entire FT can be structurally visualized on one or two pages, as opposed to the 20 or more pages in a printed FT or the large continuous FT on the computer screen.

CHAPTER 9

FTA MYTHS AND CRITICISMS

9.1 Common FTA Myths

The FTA discipline suffers from many myths or misconceptions, most of which are not quite true. These misconceptions, along with a response explaining the true reality, include the following:

1) Myth: FTA cannot model sequential timing.

    Truth: FTA can handle timing, it's just a little more difficult. The Priority AND gate and the Exclusive OR gate provide timing. In addition, multi-phase FTs provide timing coverage for system phases.

2) Myth: FTA cannot model repair.

    Truth: FTA can handle repair, however, it's a little more complex and the mathematics involved are more complex. Many computer FT programs do not support this capability.

3) Myth: FTA cannot model multiple phases.

    Truth: FTA can handle multi-phase FTs (see chapter 11.4). Drawing multi-phase FTs is more cumbersome and the mathematics involved is much more complex. In addition, many FT programs are not designed to handle multi-phasing.

4) Myth: FTA only handles independent events.

    Truth: FTA works best with independent events; however, dependent events can be modeled. The mathematics involved are a little more complex for FTs with dependent events. It should be noted that FT approximations that ignore dependency mathematics are still quite good for design decision making purposes.

5) Myth: The final FT probability calculation is an exact answer.

Truth: Quantitative FTA results are a point estimate based on the FT analyst's capability and the accuracy (goodness) of the data. A numerical result should be taken as a close approximation, but not an exact answer. A FTA does not necessarily provide 100% fidelity of a system design; it is a model or perception of reality only; it is an estimate, not an exact duplicate of the system. This is also true of most system analysis models and tools.

6) Myth: FTA is a hazard analysis tool.

Truth: FTA is primarily a root-cause analysis tool; however, it can also be used for hazard analysis. It is more limited in hazard analysis capability than other hazard analysis methods, such as Preliminary Hazard Analysis.

7) Myth: FTs can be developed using the complement of a Success Tree.

Truth: FTs can be constructed by using the complement of a Success Tree. However, for very complex systems this does not work well and important fault events are often overlooked because they do not show up in a Success Tree.

8) Myth: FTA is the same as an FMEA.

Truth: They are not the same. An FMEA is a bottom up single-thread analysis, whereas a FTA is a top down multi-thread analysis. Single-thread analysis means the analysis only looks at the single failure mode and its immediate effect. A multi-thread analysis looks at multiple event failure combinations and their effect on the system. An FMEA considers all failure modes in its analysis, while a FTA considers only those failures pertinent to the top UE. FTA and FMEA are different tools with different objectives and methodology.

9) Myth: FTA should be used to evaluate every identified hazard.

Truth: Although FTA is a powerful tool, it can also be expensive to apply. Therefore, FTA should only be applied when required by the situation. Typical reasons to perform FTA include: it's required by the customer, it's required to fully evaluate a safety-critical hazard or it's required to demonstrate compliance with a numerical requirement. FTA should not necessarily be performed on low risk hazards.

10) Myth: If two FTs of the same system look different, then one must be incorrect.

Truth: Different FT analysts often use different approaches, which result in different looking FTs, however this does not necessarily mean one is incorrect. Figure 9.1 contains two example FTs that debunk this myth. In this example, two different looking FTs are developed of the same system. Both FTS are correct and produce the same cut sets and probability calculations.

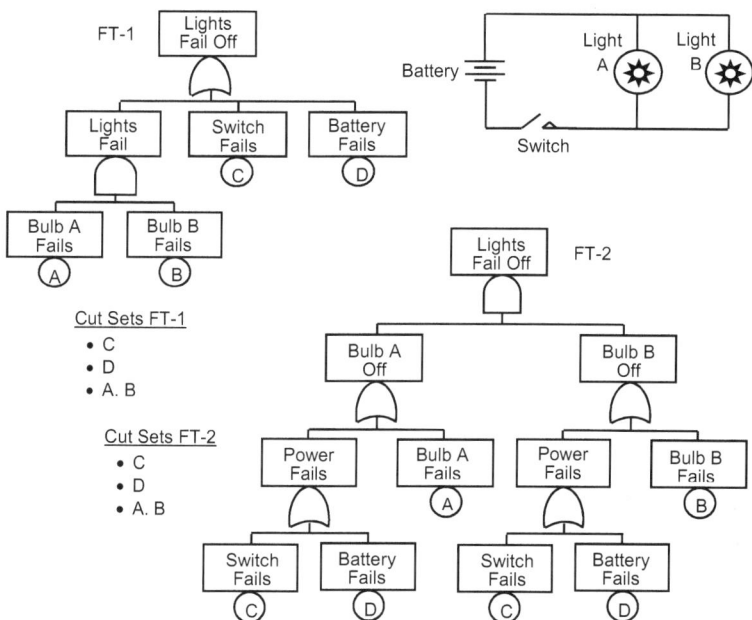

Figure 9.1 – Myth #10 Debunked

9.2 Common FTA Criticisms

FTA also suffers from several criticisms, many of which are not quite true. The following are some typical criticisms, with a response to each:

1) Criticism: It's too difficult for an outside reviewer to know if a FT is complete.

Reality: It can be difficult, particularly if the reviewer knows little about the system design. This criticism is essentially true of any analytical tool, such as Markov Analysis and Sneak Circuit Analysis. The analysis is only as good as the skill of the analyst performing the analysis; this is why an experienced FTA analyst is required. The reviewer of a FTA must have a good understanding of the system design and be somewhat familiar with FTA in order to judge the completeness of the analysis. Quite often independent FTA experts are used to evaluate the completeness and correctness of a FTA for a customer. This is recommended for complex and safety-critical systems.

2) Criticism: The correctness of a FT cannot be verified.

   Reality: This criticism is not correct; after a FT is constructed and the CSs are generated, each individual CS can be checked against the system design for correctness. If a CS cannot produce the top UE, then there is an error in the FT logic model, which means a correction in the FT is necessary.

3) Criticism: FTA failure data is questionable and therefore the quantitative results cannot be trusted.

   Reality: This criticism is partially true. In order to obtain a good estimate of the top event probability, good data is required. However, if good data is not available worst case data can be used to provide an estimate. If the top probability is acceptable with worst case data, then the system design should be acceptable even without refined data. Good data means a component failure rate has a fairly long history, which has established a known (and reliable) failure rate for the component. Worst case data means that very little history or information can be found on a component and an engineering judgment is made regarding the failure rate of the component.

4) Criticism: The uncertainty in quantitative results is unknown.

   Reality: This criticism is true because there are many variables that can affect the uncertainty in quantitative results. For example, anything left out of the FT will not show up in the final results, thus making the results an underestimate. If the failure data is mushy (has a wide range) then the results are mushy. One of the latest developments in FTA is the use of fuzzy mathematical methods in the quantitative calculations in order to obtain a range of estimated results. Fuzzy math applies a numerical range to component failure data, rather than using a point value. See references, 16, 17 and 18.

5) Criticism: Different analysts sometimes produce different looking FTs of the same system, therefore one must be incorrect.

    Reality: Different FT analysts often use different approaches, which results in different looking FTs, however this does not necessarily mean one is incorrect. The approach taken by one analyst may be different than the approach taken by a different analyst, thus making the FTs look different. The proof is in the cut sets. If each analyst did a proper analysis, the resulting cut sets should be identical. Figure 9.1 (above) is an example FT that debunks this criticism (and myth #10). In this example, two different looking FTs are developed of the same system. Both FTS are correct and produce the same cut sets and probability calculations.

6) Criticism: FTs are too subjective.

    Reality: The very top levels of large complex FTs may be somewhat subjective because this is where the analyst's approach and skill play a major role in determining how to structure the overall FT. However, the lower levels of the FT should correlate directly to the system design and architecture, not allowing any room for subjectivity. For example, a FT of a circuit diagram should follow the circuitry component-by-component. When working with FTs produced from drawings and schematics, the FT logic is very concrete and not subjective. Subjectivity only enters the FT at the top level of the FT for large systems, where the analyst must make decisions on how to structure the overall analysis.

7) Criticism: FTA is time consuming and expensive.

    Reality: FTs can easily become time consuming and expensive; the analyst must most know and understand when to stop the analysis. FTs can become large, unwieldy and time consuming if proper planning and attention is not taken. However, the cost may be worth it for safety-critical applications, where FTA is the only method of determining the true level of safety.

8) Criticism: FTA analysts require too much time developing a FT.

Reality: Sometimes FTA can become the goal rather than the solution for an analyst, because FTA has been known to totally absorb an analyst, like a video game. The analyst must stay focused on the problem and solution rather than enjoying the analysis challenge. This is why an experienced FT analyst is required.

9) Criticism: FTA requires an analyst with extensive training and practical experience.

Reality: This criticism is true of any valuable tool, including FTA. Knowing too little about FTA can be dangerous. Having a well trained and experienced FTA analyst is invaluable. A good FT analyst is typically also good at understanding systems designs. Basic FTA is easy to learn, causing the inexperienced analyst to easily get into trouble. Knowing too little about FTA can be dangerous; simplicity is seductive, making it easy to go beyond one's knowledge and understanding.

10) Criticism: No logic loops are allowed in the FT structure.

Reality: This is true, but should not be a criticism. If this were allowed, then this would cause endless loops in the FT computer analysis program. The FT can still model feedback loop situations by breaking the system loop in the FT model.

11) Criticism: FTA cannot model intermittent failures, only hard failures.

Reality: This is true, but should not be a criticism. Hard failures are the primary concern in system safety, as most intermittent failures are ironed out during system test.

12) Criticism: FTA does not always provide probabilities to six-decimal place accuracy as may be required in modeling certain system designs for reliability.

Reality: This is typically true, however comparison of FT results to those of other tools, such as Markov Analysis (MA) show that FTA provides very comparable results with much greater simplicity in modeling. And, where the failure data is usually sometimes fuzzy, approximations are adequate and useful. When dealing with extremely small numbers (i.e., 10^{-6}) decimal place accuracy does not usually help in decision making.

CHAPTER 10

FTA MISTAKES AND MISUSES

10.1 Common FTA Mistakes

When first learning how to perform a FTA it is commonplace to commit some traditional errors. The following are some typical mistakes made during the conduct of a FTA:

1) Not fully understanding the system design and operation. There is an old FTA maxim that says: if the FT analyst cannot draw a simple functional diagram of the system, then the analyst probably does not understand it well enough to develop an accurate FT.

2) Not fully understanding the FT construction process. FTA can be very seductive because the basics are very easy to learn and understand. However, this sometimes causes problems because the analyst gets in over his head.

3) Forgetting the correct FT definitions and applying the wrong gate or event type.

4) Not including human error in the FT. Anything omitted from the FT skews the final results. The human is an important component of the system design and human error is an important aspect that must be considered.

5) Jumping further ahead in the system design than the fault logic warrants. It is easy to sometimes skip ahead in the logic process when looking at the system components.

6) Not placing any text in every FT node text box. If text is missing from any node, it then becomes impossible for anyone else to follow the analyst's logic in developing the FT.

7) Not placing sufficient descriptive text in every FT node text box. The particular scenario should be adequately described. It is important to use state transition descriptions when possible, rather than just stating that something failed.

8) Failing to account for MOEs in the FT mathematics, or doing so incorrectly. If MOE generated CSs are not properly reduced by Boolean algebra the quantitative results will be incorrect.

9) Failing to perform a reality check on computer generated results to ensure the computer results are correct.

10) Developing an incomplete FT. The key to building a good FT is completeness. A complete FT must include all relevant components and fault paths contributing to the top UE. Any forgotten or missed component leads to an incomplete analysis with a potentially significant impact on the results.

11) Failing to fully understand the FT computer software being used to construct the FT. FT software must be correct in order to avoid unknown errors. Results from FT software must be checked for correctness. Questions to ask to help validate a FT computer program include:

- Are all CSs reduced to MinCSs?
- Are all CSs correctly resolved?
- Are MOEs correctly resolved?
- Is CS order truncation done correctly?
- Is CS probability truncation done correctly?
- Is the program numerically accurate?

12) Failing to test the FT computer software being used, to ensure it provides accurate cut set and probability results. When using FT computer programs to calculate the top event probability the final results should always be double-checked to ensure the results are accurate. Some computer programs have been found to have errors or use algorithms that occasionally fail under certain situations. In addition to understanding the FTA software, it is necessary to run special tests to ensure accuracy. A reasonableness calculation should always be made on the final results.

13) Failing to plan the FTA process for the project under investigation. Also, failing to give some forethought to the overall FT design and structure. Certain FT approaches may be more difficult than others. Planning decisions should be documented so that they are formally retained and remembered throughout the analysis effort.

14) Yielding to AND gate overconfidence. Sometimes when an analyst sees that an AND gate has 5 or 6 gates going into it, they rationalize that the probability will be so small it is not necessary to trace out the root-causes of the gate inputs. However, it might be possible that further down the FT structure some of the input paths might have common MOEs, thereby reducing the actual number of independent inputs to the AND gate.

15) Failing to apply the correct exposure time to failure events. Large variances in time can skew the results significantly.

16) Repeating the same text in different test boxes at different levels in the FT. For example, it is quite common to be developing a FT branch and placing the text "valve fails to open" in a text box at level 6 of the FT, and then on level 8 or 9 repeating the same words without realizing they were already used. Each text box should be uniquely descriptive without repeated text.

10.2 Common FTA Misuses

FTA is sometimes misused or abused, intentionally or unintentionally. When this happens, it hurts the reputation of the analyst and the FTA. It also opens possible ethical issues. The following are some typical misuses of FTA:

1) Manipulating the FT structure to obtain desired results. When the FTA final quantitative results do not produce the desired numerical result, there is sometimes a temptation to modify the FT to produce the desired results. If the FT correctly models the system design, then the results are showing that a design change is needed to correct the design weakness producing the undesired result. This temptation must be avoided because if carried out, the final results will be incorrect and thereby hide the true safety situation of the system.

2) Manipulating the data to obtain desired results. This misuse is similar to #1 above, except in this case the component failure rates are modified to obtain the desired numerical results. Again, this temptation must be avoided because if carried out, the final results may be significantly incorrect.

3) Sloppy and careless analysis, that results in the construction of an incorrect FT model. The FT analyst should be knowledgeable, experienced, disciplined and thorough. If the FT analyst does not put in the time and effort to understand the system and correctly and completely model it, then the final results will not be entirely correct. Errors may be detected by evaluating the FT CSs, but fault events left out of the FT will skew the final results.

4) Analyzing the wrong problem. Make sure the particular problem described by the FT top UE is the correct problem of concern.

5) Incorrect application. Make sure that FTA is the right tool for solving the particular problem under investigation.

CHAPTER 11

FTA SPECIAL TOPICS

11.1 Latency

A latent failure refers to a component that is not checked for operability before the start of a mission, thus it could unknowingly be failed when required for use. When failure of the component occurs it is not detected or annunciated. Certain latent faults can be of system safety concern when they are involved in system designs where operation is critical. Latency can significantly increase the potential safety risk because this situation effectively increases the component exposure time. The latent time period is the time between maintenance checks, which can often be significantly greater than the mission time. This large exposure time can make a large impact on the probability of the UE. Latency is also sometime referred to as a dormant failure.

Figure 11.1 provides an example FT showing how significant the numerical error can be if latency is not properly accounted for in the FTA. In this example, the FTA models an uncontained fire in the lower bay of a commercial aircraft. There is a fire detection/suppression system in the bay to detect and extinguish a fire. An uncontained fire will result if the fire detection/suppression fails, causing loss of aircraft. The FT on the left evaluates the design ignoring the effect of latency. It assumes both components are checked for failed state prior to flight. However, in reality the fire detection/suppression may be checked only every 6,000 hours (contrived number) because it can only be done in the maintenance shop. The FT on the right shows the correct calculation which accounts for the fact that the fire detection/suppression system cannot be checked for failed state prior to flight, thus increasing the effective exposure time. Note the two orders of magnitude difference in the probability calculations, and that the FT ignoring latency understates the actual risk.

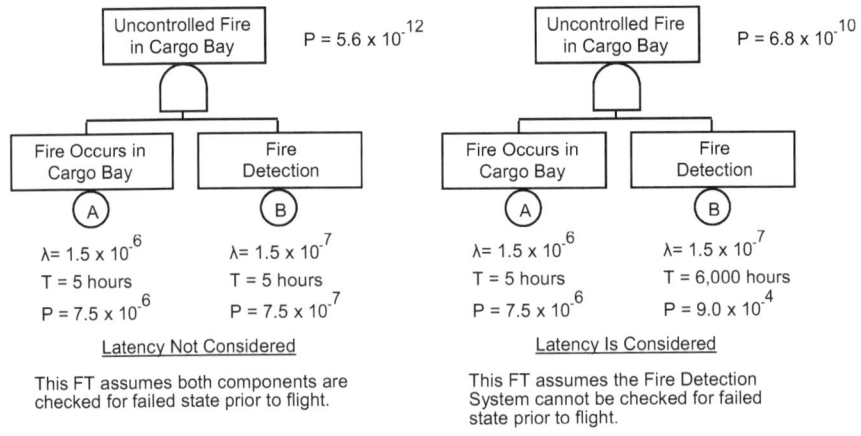

Figure 11.1 – FTA Example of Latency

11.2 Common Cause Failure

A Common Cause Failure (CCF) is the simultaneous failure of multiple components, typically redundant components, due to a common or shared cause. For example, a CCF occurs when two redundant electrical motors become inoperable simultaneously due to a common circuit breaker failure that provides power to both motors. In this example the common circuit breaker provides the CCF event. CCFs can include causes other than just design dependencies, such as environmental factors, human error, etc. Ignoring the effects of CCFs in the system design can result in overestimation of the level of safety and reliability (i.e., the level of safety appears better than it actually is). CCF is an insidious method of negating independent redundant designs.

CCFs create the subtlest type of hazards because they are not always obvious, making them difficult to identify. The potential for this type of event exists in any system design that relies on redundancy or uses identical components or software in multiple subsystems. CCF vulnerability results from system failure dependencies inadvertently designed into the system. CCFs can be caused from a variety of sources or coupling factors, such as:

- Common weakness in design redundancy, such as close proximity of redundant safety-critical hydraulic lines
- The use of identical components in multiple critical subsystems
- Common software design

- Common manufacturing errors
- Common requirements errors
- Common production process errors
- Common maintenance errors
- Common installation errors
- Common environmental factor vulnerabilities

Figure 11.2 contains two FTs for an example 3-redundant element system; one FT without consideration of CCF and a second FT that includes a CCF event. This example system consists of three redundant motors, where system failure results when all three motors have failed. A CCF can cause all three Motors to fail simultaneously (P_{ABC}), thereby bypassing the intended design redundancy. The probability of P_{ABC} is the critical factor.

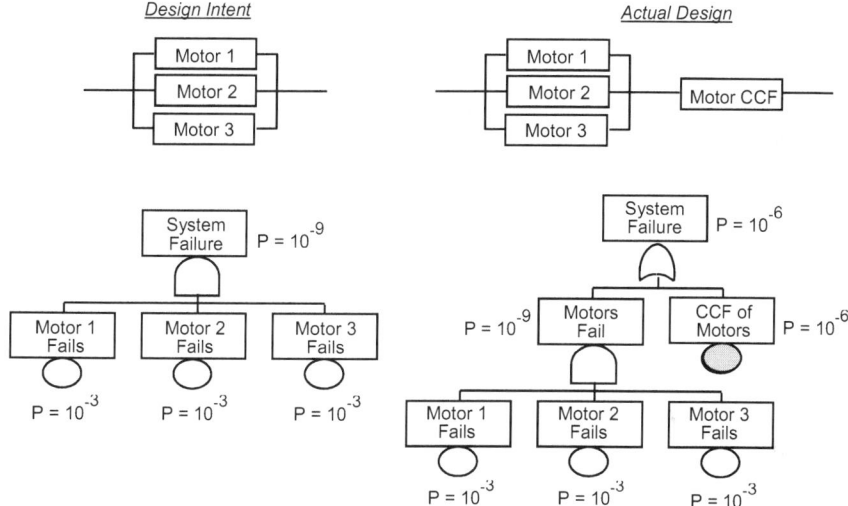

Figure 11.2 – FTA Example of CCF

Note that the initial FT produces a 3-order AND gate that results in a system probability of failure of $P = 1.0 \times 10^{-9}$. The updated FT that includes CCF produces a probability of system failure of $P = 1.0 \times 10^{-6}$. This example demonstrates the impact of a CCF, which hinges on the probability of the CCF event.

11.3 Interlocks

An interlock is a design safety arrangement whereby the operation of one control or mechanism allows, or prevents, the operation of another function. The safety interlock is a special safety device used in a system design to increase the level of safety of a specific function; it is a design safety feature. The primary purpose of an interlock is to provide a mechanism to make or break a safety-related function, based upon a set of pre-determined safety criteria. It should be noted that interlocks are not necessary for the operational functionality of a system. An interlocks is a device added to the design in order to achieve the needed safety required of the system, not for the operational effectiveness of the system. Interlocks control state transitions in an attempt to prevent the system from entering an unsafe state or to assist in exiting from an unsafe state. Interlocks can be in various forms, such as: switches, relays, barricades, shields, lockouts, procedures, logic, or walls. Interlock weaknesses include: CCFs, dependency between interlocks (software for example), human error, or bypasses (such as a Battle Short).

An Interlock can be used in either of two ways: (a) to break a function when necessary, and (b) to make a function complete when necessary. The break-function interlock interrupts a critical system function when a known hazardous state is about to be entered, thereby preventing a mishap. For example, if a hazardous laser is being operated in a locked room with no personnel in the room, a sensor-switch interlock on the door automatically removes, or breaks, power from the laser system when someone inadvertently entering the room opens the door. This interlock breaks the laser operation function and prevents a potential eye injury mishap. The system cannot be initiated and operated unless the "door is closed" safety criteria is satisfied. In this case, an unsafe system state is prevented by forcing the system into a known safe state (i.e., laser off state). Once the door is opened, the system must be re-initialized, with assurance that no one is in the room. In another example, a break-function interlock could be used to automatically remove electrical power from contacts inside an electrical panel when the panel door is opened, thereby eliminating the possibility of accidental personnel electrocution.

The make-function interlock prevents a safety-related function from being executed until the interlock safety criteria become valid. For example, in a certain missile system design, the power to launch a missile would not reach the missile until three separate and independent switches are intentionally closed. The switches are interlocks and switch closures are based on timing and passing the safety criteria at each point in time. Each switch requires certain safety criteria to be true before the switch is closed. This type of safety interlock protects against inadvertent or premature function that could result from possible system failure modes.

Figure 11.3 provides a pictorial representation of a missile launch interlock example using three switches in the functional launch path, along with its corresponding FT. These switches are considered as three independent interlocks that significantly reduce the probability of missile launch power inadvertently reaching the missile due to random failures in the system. The hazardous launch state cannot be entered until all three interlocks have passed their safety criteria and closed. The decision logic box represents the methodology selected for monitoring and judging the safety criteria. It is important to ensure that there is valid justification for each interlock, and that there is a suitable safety case showing that the interlocks are completely independent and not subject to common mode or common cause failures. It is also important to ensure that the combined probability of failure for the three interlocks is adequately small; if parts with high failure rates are used the safety advantage of the design may be negated by an overall high probability of activation. This interlock FT illustrates how the interlock design protects against single point failures. It also shows how the design introduces AND gates into the fault tree logic, which help to reduce the overall probability of the UE.

For normal system operation, missile launch would be expected when all of the following conditions are successful: a) launch power is made available, b) interlock-1 closure occurs, c) interlock-2 closure occurs AND d) interlock-3 closure occurs. The FT models these normal intended events occurring prematurely or inadvertently due to system failures.

In Figure 11.3 the input events are shown as Diamonds to indicate that they can be further expanded via FT logic when the design is analyzed in more detail. It should be noted from this FT that if the interlocks were not in the system design, inadvertent launch would occur as soon as event A occurred, and the resulting top probability would be about 1.0×10^{-3} as opposed to the 1.0×10^{-12} with the three interlocks. It is clear from the FT that the top probability is directly affected by the numbers of interlocks utilized, as shown in Table 11.1. Changes in the individual interlock probabilities will also influence the top probability.

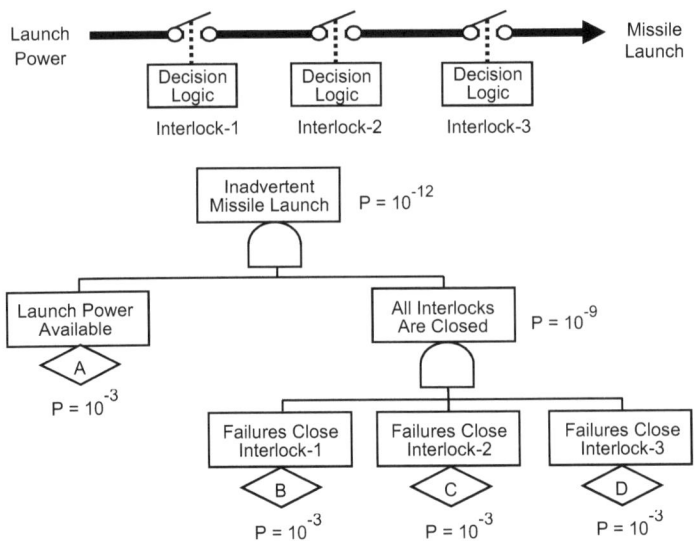

Figure 11.3 – FTA Example of Interlock

Table 11.1 summarizes the top probability based on the number of interlocks in this example system: 0, 1, 2 or 3. Obviously, the more interlocks in the functional path the lower the probability of inadvertent functioning.

Table 11.1 – Interlock Influence on Fault Tree Top Probability

Number of Interlocks	Top Probability
0	$P = 10^{-3}$
1	$P = 10^{-6}$
2	$P = 10^{-9}$
3	$P = 10^{-12}$

11.4 Multi-phase FTA

Most FTs are a single phase FT (SPFT), meaning they represent a single phase of system operation. If a system has multiple phases of operation a SPFT is drawn for each individual phase. Multiple SPFTs may result in considerable duplication of FT branches. In order to avoid duplication of FT structure and analysis effort, an alternative approach is to use a multi-phase fault tree (MPFT). In a MPFT multiple system phases are handled in a single FT structure, as opposed to multiple FT structures.

In a sense, a MPFT also handles timing and sequencing events that occur during phase changes. The problem with MPFTs is that the FT data handling and mathematics becomes much more complex. For this reason, many FT computer programs only handle SPFTs.

Figure 11.4 demonstrates the concept of multiple SPFTs for multiple system phases. Note that some events occur in one or more phases, while other events occur only in a particular phase.

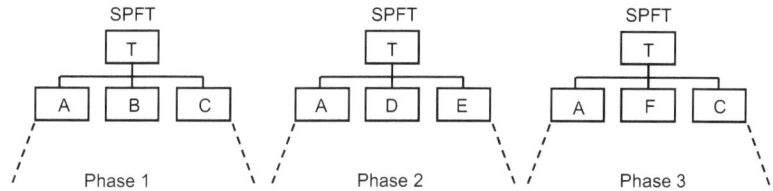

Figure 11.4 – Three SPFTs for Three Phases of System Operation

Figure 11.5 demonstrates the concept of one MPFT for several system phases. This approach requires a phase table to keep track of the phases in which each event occurs.

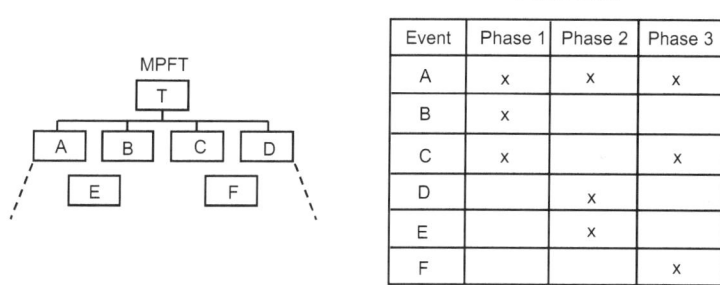

Phase Table

Event	Phase 1	Phase 2	Phase 3
A	x	x	x
B	x		
C	x		x
D		x	
E		x	
F			x

Figure 11.5 – MPFT for Three System Phases

There are various reasons necessitating the need to model several system phases. Various system factors can change from phase to phase. For example:

- The system configuration may change in different phases
- Operator tasks may impact different phases
- System safeguards could be intentionally removed during a specific phase

- Repair may take place during a specific phase
- Specific system tests could take place in a certain phase
- Failure rates can change in different phases due to stress, temperature, etc.

Complex systems generally have more than one phase of system operation. Figure 11.6 depicts the various phases of operation for a commercial aircraft. The length of time for each phases is different, as well as the system variables and states.

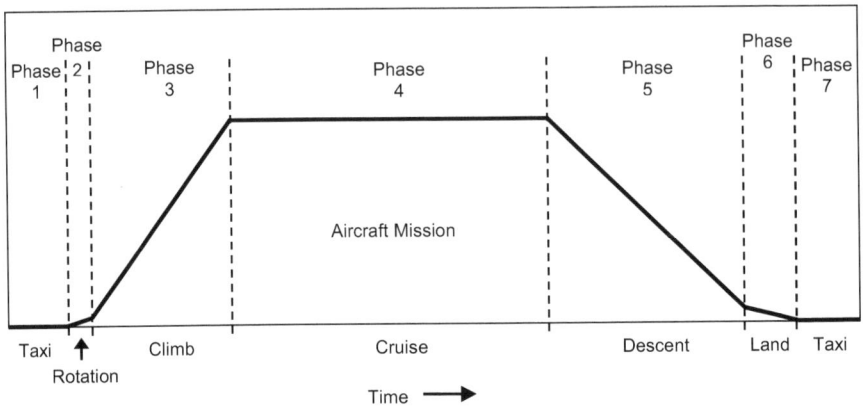

Figure 11.6 – System Phases for a Commercial Aircraft

Table 11.2 shows some of the system variables that can change during the different mission phases of the commercial aircraft. These variables could have a significant impact on the FTA and the safety of the system. If a MPFT were not employed in this situation, a separate SPFT would have to be developed for each individual phase. This table provides a small example of the many changes that take place during an aircraft mission, each of which must be considered for safety impact.

Table 11.2 – Aircraft System Changes during Mission Phases

Mission Phase	System Changes
Climb	• Landing gear is retracted • Squat switches are opened or closed to enable certain functions • Component stress is slightly higher • Rapid changes in temperature, pressure and humidity occur
Cruise	• Autopilot is engaged • Navigation system is engaged
Descent	• Landing gear is extended • Rapid changes in temperature, pressure and humidity occur
Taxi	• System power cycles through off-on-off sequences • Squat switches are opened or closed to prevent certain functions

11.5 Redundancy

Redundancy is both a design safety feature as well as a design reliability feature. Redundancy involves the duplication of critical system components with the intention of increasing reliability of the system, or the system function which the redundant elements are performing. When a single component or system element does not produce the desired level of reliability, then two or more of these elements are linked together in parallel, thereby significantly increasing the reliability. This concept is demonstrated in Figure 11.7.

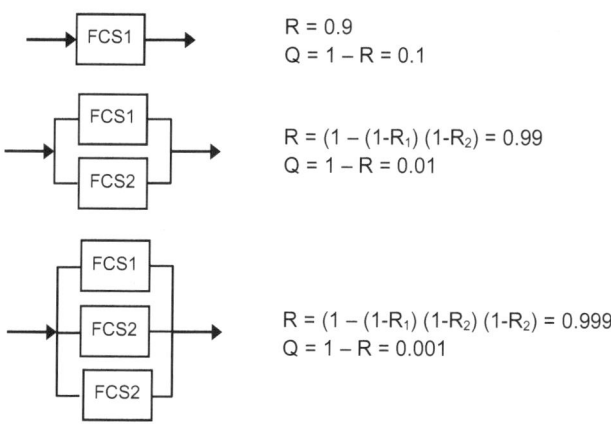

$R = 0.9$
$Q = 1 - R = 0.1$

$R = (1 - (1-R_1)(1-R_2)) = 0.99$
$Q = 1 - R = 0.01$

$R = (1 - (1-R_1)(1-R_2)(1-R_2)) = 0.999$
$Q = 1 - R = 0.001$

Figure 11.7 – Redundancy Concept

In this simple example an aircraft has a Flight Control System (FCS) unit with a reliability R = 0.9. This means the unreliability (Q), or probability of failure is P_F = 0.1. For safety this probability of failure is too high and unacceptable. Therefore, two FCSs are placed in parallel, thereby increasing reliability to R = 0.99 and the probability of failure is now P_F = 0.01, which is still not acceptable. When three FCSs are placed in parallel, R= 0.999 and P_F = 0.001, which is now acceptable. In this configuration reliability and safety have improved because system failure requires loss of all three FCSs (as shown by the AND gate). In many safety-critical systems, such as aircraft fly-by-wire and hydraulic systems, some parts of the control system may involve triplex redundancy in order to ensure maximum reliability and safety.

This triple redundant design gives great confidence that the system design is now safe. However, to really ensure this assumption is correct, a detailed FTA of the design should be performed to ensure that there are no common-cause failure (CCF) modes that could eliminate or bypass the redundancy. An example of this situation is demonstrated in Figure 11.8. where it is assumed that the triple redundant FCS has been analyzed in detail, and after going down 10 levels in the system design it was discovered that Power Supply A (PS-A) provided power to each of the separate FCS units. One of the many CSs that would be generated for this hypothetical system would be event A, which becomes a single point failure that bypasses the redundant design. It is important in FTA to continue analyzing the system until the FT reaches the basic component level. There is always a temptation to stop analyzing when a 3-input AND is reached at a high system level.

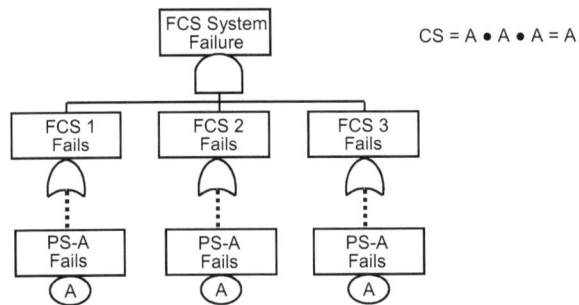

Figure 11.8 – Redundancy Bypassed by CCF

Redundancy is very useful, but it can also be tricky. As shown in the above example, CCFs are always a concern with the use of redundancy. In addition, sometimes redundancy works against safety depending on the system usage and the UE involved. Figure 11.9 demonstrates an example case where redundancy is not beneficial to safety.

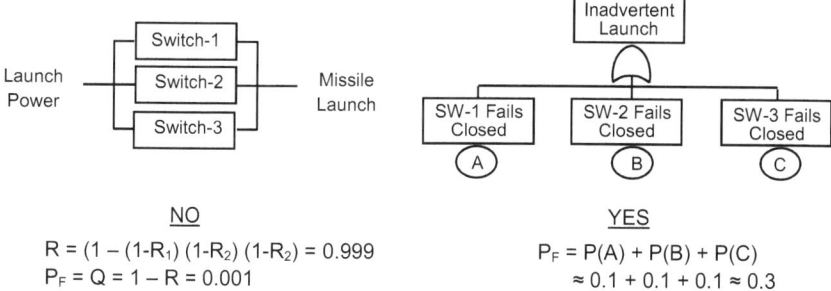

Figure 11.9 – Redundancy Can Hurt Safety

In this simple example, when the switch is closed by the operator, power is allowed to go to a missile launch initiation component that initiates launch. A switch typically has two failure modes: open and closed. The switch failure probability would be allocated between these two modes. Safety is concerned about the switch failing closed because this mode would automatically and unintentionally launch the missile. In this situation safety is concerned about the UE "inadvertent launch" and the fail closed mode of the switches. One might be tempted to use Q = 0.001 calculated from the unreliability of the triple redundant switches. However, in this case that would be the incorrect approach. In this situation the UE must be calculated using an OR gate, because if any one of the switches fails the UE occurs. For cases like this the thought process changes from unreliability to *undesired* event. Because of the redundancy there are three opportunities to cause the undesired event rather than just one.

11.6 Accident Investigation FTA

FTA is normally a *proactive* analysis tool for predicting potential causes of postulated UEs during the design of a new system. It is also a *reactive* analysis tool for ferreting out the root-causes leading to an UE that has actually occurred, such as an accident. FTA is very powerful as a structured methodology for identifying root-causes, plus it also provides a visual communication model that most individuals can readily understand and follow. The visual model displays the logical progression in the chain of events leading to an accident.

In an accident investigation it is important to efficiently identify the root-causes of an accident. In order to increase efficiency and effectiveness, accident data can be used in the FTA to identify which paths to follow and which paths are not in the causal path chain.

The Evidence Event Fault Tree Analysis (EEFTA) provides a methodology for incorporating evidence into both the analysis and the FT model. It indicates where evidence is needed in order to make a decision regarding root-causes. At the same time, the evidence closes analysis effort on paths that did not cause the incident. These paths can be re-opened at a later time if necessary.

EEFTA utilizes the Evidence Gate (EG) to either open or close an analysis path depending upon the evidence. The EG opens a FT branch when there is evidence to support it, or lack of evidence sufficient to make a judgment, and it closes a FT branch when there is sufficient evidence to indicate that particular path did not contribute to the accident. This methodology allows the analysts to consider all possibilities and to show why certain paths were ruled out.

When evidence is available to show that a particular branch did not happen, then that branch is ruled out and no further analyses need progress down that path. If there is no evidence, or positive evidence supporting a path, then this indicates the direction the analysis should follow. The FT model shows what was considered and ruled out, and why certain paths were followed, hopefully to the actual root-causes. The FT model considers all possible contributors, such as hardware failures, software errors, personnel errors, environmental factors, procedure errors, etc.

The methodology for developing an accident investigation FTA is very similar to normal FT construction. There are two major steps involved:

1) First Pass FT – Analyze system and incident using normal FTA construction rules and system logic. Identify and establish major system fault states that could possibly lead to the accident.

2) Second Pass FT – Go through the first tree and determine where known evidence applies, or where additional evidence is needed. Place these conditions in the FT using the Evidence Gate. Continue down branches with positive evidence or insufficient evidence. Terminate branches with negative indicating evidence.

Figure 11.10 shows the structure of the Evidence gate. As shown in this diagram, the evidence condition is attached to the right side of the gate. The collected empirical evidence is placed inside this conditional event node. The evidence supports the output node either positively or negatively. If the conditional event node is true the analysis continues, if false the analysis stops at that point. Occasionally a fault tree will be developed either independent of the evidential data, or it may be developed faster than the data is collected. In these cases the FTA may go past the point where evidence would terminate a branch. The FTA is also instrumental in guiding the investigation in terms of where and what evidence data is needed.

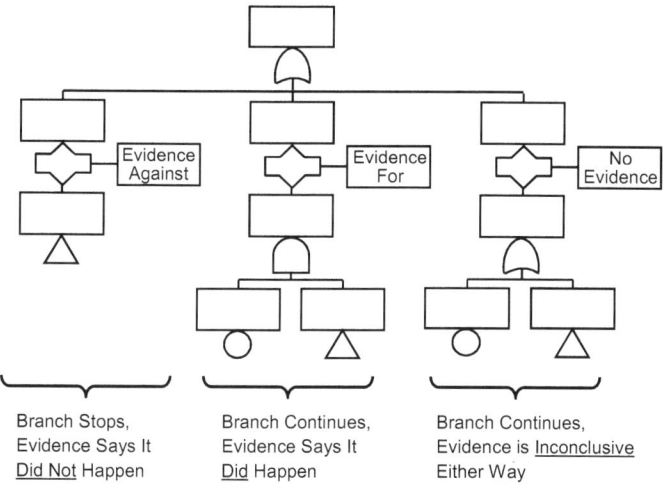

Figure 11.10 – Evidence Gate in Accident Investigation FTA

Accident investigation FTs tend to typically be smaller than design development fault trees. This is because in the proactive design phase FT all possible branches are analyzed to their root-causes. In the accident FT only those branches contributing to the accident are carried through to their root-causes.

11.7 FTA of Software

FTA of both hardware and software is possible; however, it is more effective on hardware than software. Performing FTA on software code is somewhat futile because software code does not have identifiable failure modes. However, software functions do have failure modes that can be identified and utilized in a system FTA. This means that software is best approached from a *system* oriented FTA that starts at the hardware level and works down through the system until the software functional entry points are reached. This approach will identify safety-critical software modules that can then be evaluated in more detail using another safety analysis technique.

In order to demonstrate this concept, a FTA is provided here on an example aircraft landing gear system. Figure 11.11 contains a simplified diagram of this landing gear system design, along with the software code module for lowering the landing gear. A hazard analysis has identified that "failure to lower the landing gear for landing" is a safety-critical UE. Figure 11.12 contains a system FT for this UE.

In this example system, the GDnB button is pressed to lower the gear, and the GupB button to raise the gear. When the gear is up, sensor switch S1 sends a True signal to the computer, else it sends a False. When the gear is down, sensor switch S2 sends a True signal to the computer, else it sends a False. When the gear is up, then S1=T and S2=F. If the gear is down, then S1=F and S2=T.

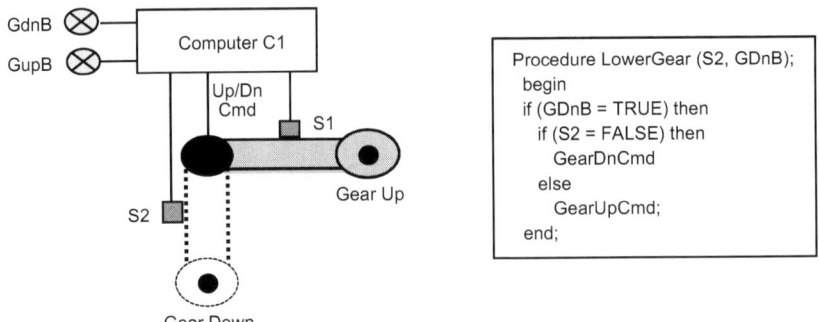

Figure 11.11 – Landing Gear System

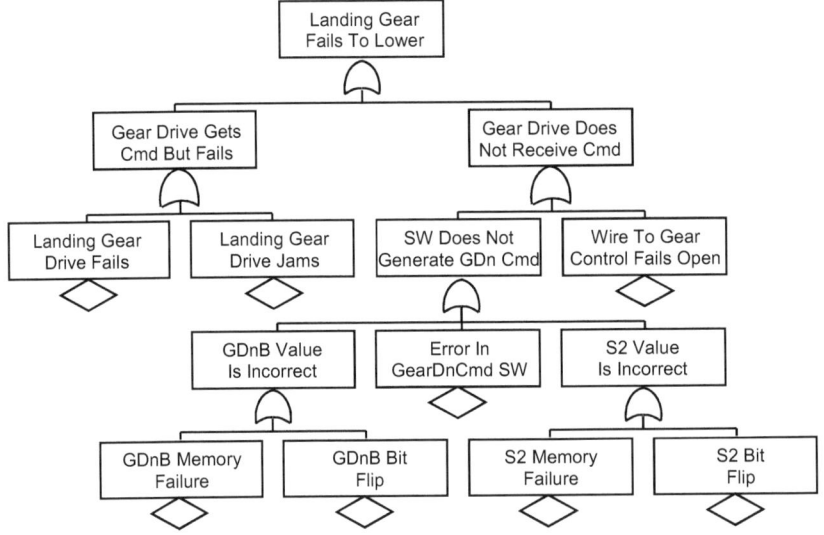

Figure 11.12 – Landing Gear System FTA

There are several important conclusions to draw from this FTA: 1) two computer memory faults affect critical software variables, 2) the GearDnCmd software module is also safety-critical and should be investigated and 3) the FTA shows all single-point failures causing the UE, thus the software design could be improved upon to eliminate single-point failures.

CHAPTER 12

FTA RELATED TECHNIQUES

12.1 Success Tree

Success Tree Analysis (STA) is the complement of FTA; it is used to calculate the probability of success as opposed to the probability of failure. In a success tree (ST) the logic gates are inverted from those of the corresponding FT, for example a FT OR gate becomes a ST AND gate. Some of the key event words are also inverted, for example, *failed* in a FT becomes *success* in a ST. For small systems a ST can be derived by inverting a FT, however, for large systems this technique does not always work correctly. Figure 12.1 shows a ST and FT for the same system.

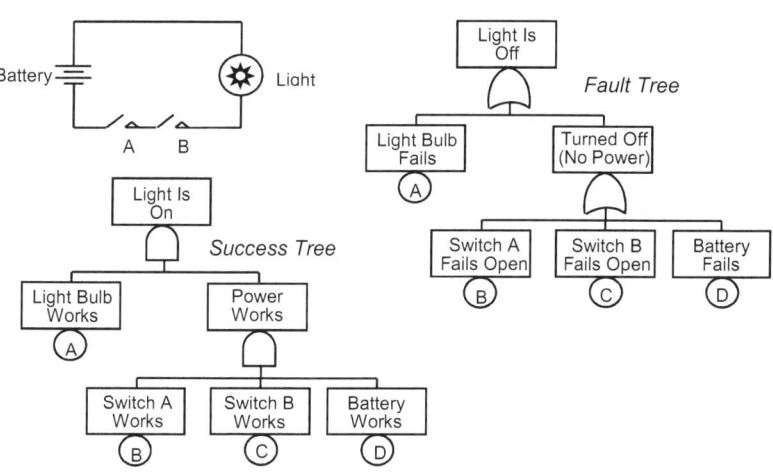

Figure 12.1 – Success Tree vs. Fault Tree

12.2 Event Tree

Event Tree Analysis (ETA) is an analysis technique for evaluating the potential outcomes following the occurrence of an initiating event. ETA utilizes a visual logic tree structure known as an Event Tree (ET). The objective of ETA is to determine whether the initiating event will develop into a serious mishap, or if the event is sufficiently controlled by the safety systems and procedures implemented in the system design. An ETA can result in many different possible outcomes from a single initiating event (IE), and it provides the capability to obtain a probability for each outcome. The results lead to recommendations as to whether or not added design safety features are needed. ETA is based upon the event scenario concept shown in Figure 12.2. This methodology evaluates the effect of various pivotal events following the occurrence of an initiating event. The pivotal events may represent the success or failure of various design safety features. An accident scenario contains an IE and one or more pivotal events leading to an end state as shown in Figure 12.2. An ET is essentially a binary form of the decision tree, where the branches are either success or failure.

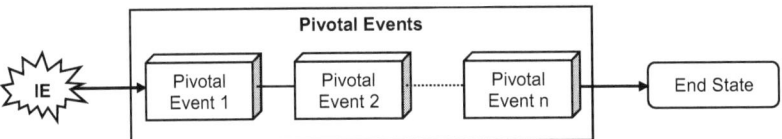

Figure 12.2 – Event Scenario Concept

The ETA begins with the identified IE listed at the left side of the diagram, as shown in Figure 12.3. Design safety methods or countermeasures are then listed at the top of the diagram as pivotal events. Each pivotal event is evaluated for successful operation and failing to operate. The resulting diagram combines all of the various success/failure event combinations and fans out to the right in a sideways tree structure. Each success/failure event can be assigned a probability of occurrence, and the final outcome probability is the product of the event probabilities along a particular path. Note that the final outcomes can range from safe to catastrophic, depending upon the chain of events involved.

An ET distills the design into a tree structure that is used to help classify potential scenarios and their consequences. Each distinct path through the ET is a distinct scenario. By informal convention, where pivotal events are used to specify system success or failure, the *lower* branch path is considered to be the *failure* path and the *upper* branch the *success* path. ETA typically utilizes a FTA to determine the causes and probability of a particular pivotal event failure.

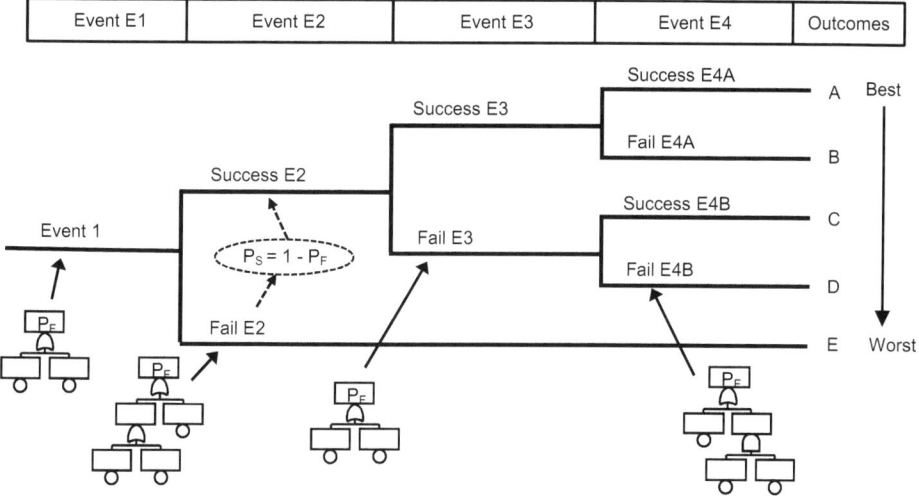

Figure 12.3 – Event Tree Diagram

The probabilities for each of the outcomes can be derived by multiplying the event probabilities in each path. For example:

Outcome A: $P_A = P_F(E1) \times P_S(E2) \times P_S(E3) \times P_S(E4A)$
Outcome B: $P_B = P_F(E1) \times P_S(E2) \times P_S(E3) \times P_F(E4A)$
Outcome C: $P_C = P_F(E1) \times P_S(E2) \times P_F(E3) \times P_S(E4B)$
Outcome D: $P_D = P_F(E1) \times P_S(E2) \times P_F(E3) \times P_F(E4B)$
Outcome E: $P_E = P_F(E1) \times P_F(E2)$

12.3 Logic Tree

A logic tree diagram is a breakdown of a problem by logic and natural hierarchy levels; it represents a logical and consistent decomposition of a problem into separate elements. For example, a problem can be decomposed structurally by constituent parts classification, such as size, shape, color or age. A problem can also be decomposed by essential relationships, such as stakeholders, objectives or goals. The overall process involves:

- Break down the problem into separate elements
- Establish groupings, priorities, classes, etc.
- Rank the elements by relative importance
- Combine relationships and sub-groups

This technique is probably the simplest of the tree models, as shown by an example in Figure 12.4. It is a good tool for group participation and brainstorming as hierarchies are a fundamental human thought process. This tool helps ensure the elements are grouped logically and consistently.

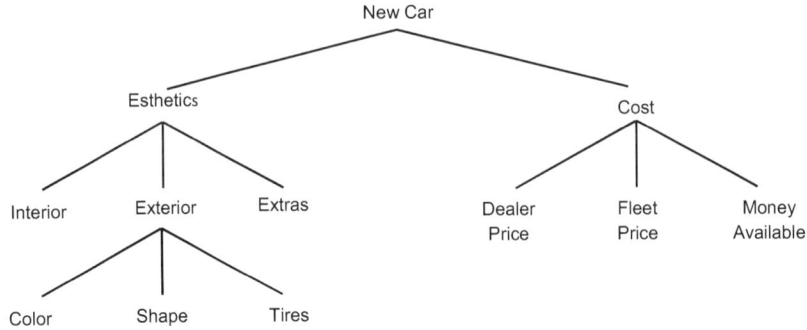

Figure 12.4 – Logic Tree Diagram

12.4 Mishap Tree

A Mishap Tree is a structured diagram that organizes Top Level Mishaps (TLMs) into an easy to visualize and understand pattern. A TLM is a generic mishap category for collecting and correlating related hazards that share the same general type of mishap event outcome. A TLM is a common mishap outcome caused by one or more hazards; its purpose is to serve as a collection point for the potential hazards that can result in the same overall TLM outcome, but have different causal factors.

TLM is the term used for a categorization method developed to provide focus and perspective on a safety issue, while also helping to simplify the tracking and classification of many individual hazards. When several different hazards can result in the same type of overall mishap outcome, that mishap event can be classified as a TLM for the purpose of correlating related hazards and risk levels. The TLM becomes a generic mishap category for collecting various hazards contributing to the particular outcome category and provides a design safety focal point for that particular safety concern.

TLMs help highlight and track major safety concerns and provide a design safety focal point. "Top Level" implies an inherent level of safety importance, particularly for risk visibility at the system level for a risk acceptance authority. The TLM severity will be the same as that of the highest contributing hazard's severity. Most hazards within the TLM will have the same severity level as the TLM; however, some may have a lesser severity. Figure 12.5 shows the generic concept of a Mishap Tree.

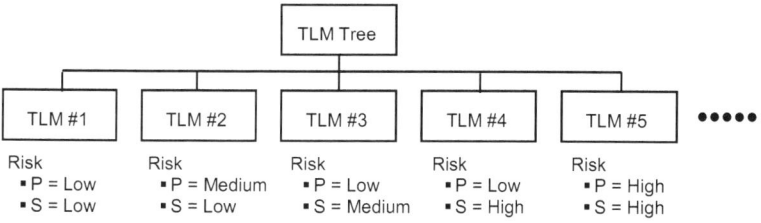

Figure 12.5 – Mishap Tree Concept

TLMs are derived by extracting the significant and common generic outcome event portion of the contributing hazards mishap description. There are significant advantages in utilizing the TLM categorization process. It groups similar hazard, while also grouping similar risk categories. Although it's not realistic to sum the risk of all identified system hazards, it is feasible to sum the risk within TLM categories. The risk presented by all hazards cannot be summed because it is not meaningful to sum hazards with significantly different severity categories.

The TLM focuses on just the significant portion of the outcome described within a hazard. The TLM is restated as a simple generic outcome event when the TLM wording is shortened or slightly revised in order to make it a generic statement. TLM wording should focus on the particular safety issue of concern. For example, the following is a list of TLMs that would form a Mishap Tree for a missile system:

- Inadvertent missile launch
- Inadvertent warhead initiation
- Incorrect missile target
- Inability to destroy errant test missile
- Personnel electrical injuries
- Personnel mechanical injuries
- Personnel RF radiation injuries
- Weapon-ship fratricide

As shown if Figure 12.6, the Mishap Tree looks strikingly similar to a FT. This Mishap Tree was adopted from the PLOA FT discussed back in Figure 6.7. As can be seen, it shows the major mishap categories, each of which has a different likelihood and severity level.

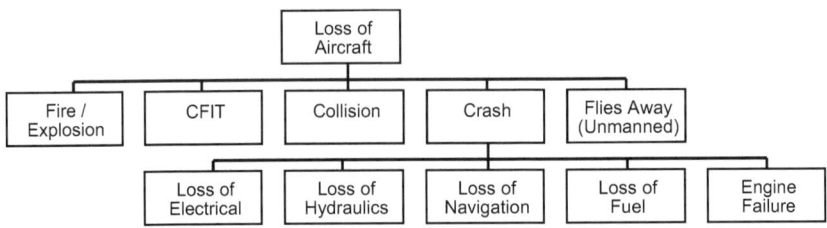

Figure 12.6 – Aircraft System Mishap Tree

12.5 Decision Tree

A decision tree is a multi-branch diagram that analyzes various options to a problem. Numerical factors can be applied to the branches in order to determine the various quantitative outcomes. The event tree is a binary form of the decision tree, oriented to pass/ fail branches and probabilities, whereas the decision tree provides more options for each branch. Figure 12.7 provides an example of a decision tree analysis for two potential investments.

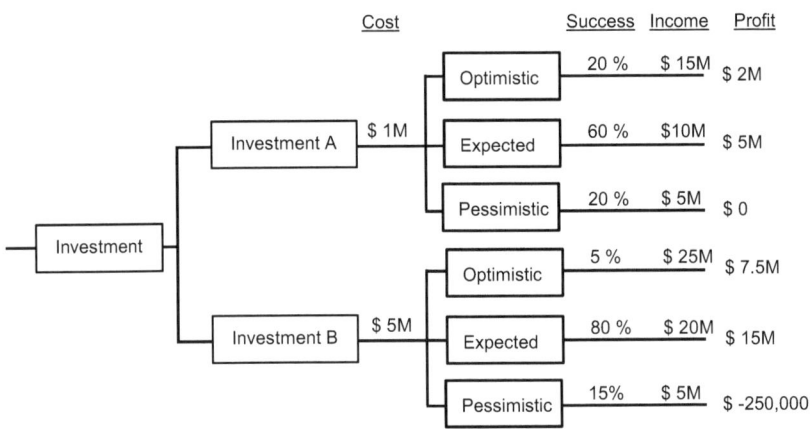

Figure 12.7 – Decision Tree Diagram

CHAPTER 13
FTA SOFTWARE

13.1 FTA Software Design Considerations

When FTA was first developed, there were no commercial computer programs available to draw and quantitatively evaluate FTs. At that time private companies developed their own proprietary in-house software that ran on mainframe computers. That situation, however, has changed for the better. Today there are numerous commercial FTA software programs available that run on desktop and laptop computers.

Computer software for FTA, referred to as *FT software*, is a valuable tool for the FT analyst. FT software is a specialized software drawing program that is similar in nature to a word processor program. FT software provides a means for a FT analyst to draw FTs, makes changes to FTs, to save and store FTs, and to produce nice looking FT hardcopies that are output on printers and plotters. FT software typically provides the following basic functional capabilities:

- FT Construction
    - Create FTs
    - Edit FTs
    - Save FTs in files
- FT Evaluation
    - Generate cut sets
    - Generate probability calculations
    - Generate importance measures
- Print FTs
    - Generate quality multi-page prints
    - Provide a means to import FTs into documents

- Generate FT reports
    - Results
    - FT data

Figure 13.1 shows the input, output and functions that a typical FTA software program performs. There are six major functions of a FT computer software code, as shown in Figure 13.1.

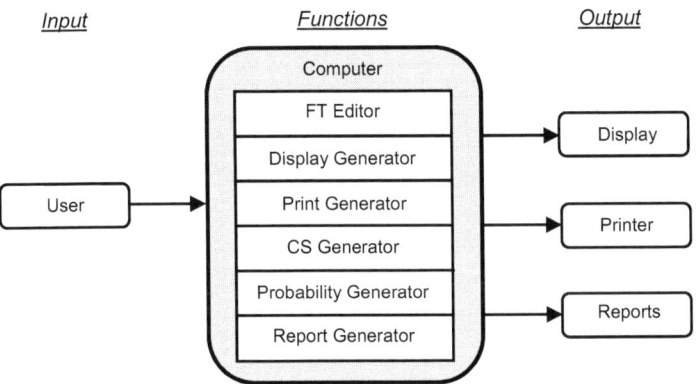

Figure 13.1 – Overall FT Software Design

13.2 FT Software Purchase Considerations

When purchasing a FT program there are three important considerations: features, price and flexibility. Table 13.1 lists some typical feature that should be considered. These features are divided into two categories: 1) features that are absolutely required, and 2) features that are not necessarily required, but are desired or nice to have. These features also indicate the flexibility of the program.

Table 13.1 – FTA Software Features

Property	Required	Desired
Construction (create, edit, copy, paste, save)	X	
Generate Min CS	X	
Generate probabilities (top, gates)	X	
Employ CS cutoff methods (order, probability)		X
Numerical accuracy	X	
Generate reports	X	
Detect tree logic loops	X	
Data export capability (tree structure, failure data)		X
Graphic export capability (BMP, JPG)		X
Tree pagination for prints	X	
User selectable print size		X
Program that does not require special installation		X
Notes on FT Print/Plot		X
Find feature		X
Undo feature		X
Employ MOEs / MOBs	X	
Correctly resolve MOEs / MOBs	X	
Global data change (failure rates, exposure time)		X
Print/Plot selected pages or all pages	X	
Unrestricted FT tree size	X	
Automatic naming of gates / events when created	X	
Capability to resolve large complex FTs	X	
User friendly (intuitive commands)	X	
Print/Plot results that are visually aesthetic		X
Generate importance measures		X
Open data file structure (i.e., can be openly read)	X	

When purchasing a FT program there are typically different options available, and vendors price their options differently. The cost of these options can vary significantly between vendors, so the FT analyst should shop and compare wisely. The following are some commercial software purchase options that must be carefully evaluated:

- Basic price (one time or yearly)
- Ownership or lease
- Maintenance cost
- Single user seat vs. multiple user seats
- Cost to change user seats
- Training costs

- Customer support cost
- Single computer or network versions

As with most software packages on the market today, some are good, some are mediocre and some are possibly bad. In addition, some programs are user-friendly and some are extremely difficult to use. For these reasons it is incumbent on the FT analyst to fully understand a FT program before purchasing. It is also necessary for the FT analyst to fully comprehend the program in order to use it optimally.

The following are some factors the FT analyst should understand about the program he/she is using:

- FT limits and capabilities
    - Max tree size
    - Max cut set size
    - Max print size
- Numerical accuracy
- Ease of use (without tech support)
- Single phase vs. multi-phase methods
- Evaluation algorithms and approximation methods
- Approximations and cutoff methods
- Gate probability calculations when MOEs are involved

Before buying a FT program it is important to take it for a test drive (i.e., test the tool). Testing will demonstrate the program's capabilities, correctness and ease of use.

Also, do not assume quantitative answers from the FT program are always correct. Depending upon the algorithms and approximation methods used, the program may not provide correct results for certain FT layouts. For this reason, it is important to always perform a manual reasonableness check (sanity check) on the results that are produced.

CHAPTER 14

FTA DATA

14.1 Failure Data

FT quantification requires a probability of failure value for all basic failure events in the FT. For example, a probability of failure (P_F) must be obtained for the FT failure event "relay fails open". Event probability is calculated using *failure data* that is directly related to the particular component, assembly or subsystem. The probability of a component failure is calculated using the equation $P_F = 1 - e^{-\lambda T}$, where λ is the failure rate for the item and T is the exposure time for the item.

Failure data can come in the form of a frequency or a probability, depending on how it is obtained. If the failure data comes in the form of a probability, then no calculation is necessary and it can be directly used in the FTA evaluation. If it comes in the form of a failure rate frequency, then a calculation is necessary, using the above equation.

All components have one or more failure modes and failure rates associated with those modes. For example, a resistor can fail in the modes of: open, shorted or out of tolerance. Failure data can either be in the form of a prediction or historical field data. A predicted failure rate is one that has been predicted from various methods, such as analysis, test or prediction formulas. A predicted rate is used when there is no historical information or data on an item. A historical failure rate is one that has been derived from actual field use. As the item has more operational hours on it, the historical rate will approach its true failure rate value and is therefore the most useful.

14.2 Failure Rate

Reliability is typically defined as "the probability that a device will perform its intended function, without failure, during a specified period of time under stated conditions". Reliability relies heavily on statistics, probability theory and reliability theory. The function of reliability engineering is to develop the reliability requirements for a product, establish an adequate reliability program, and perform appropriate analyses and tasks to ensure the product will meet its requirements. The failure rate of an item is a key value in reliability engineering.

In reliability theory, R is the probability of successful operation of an item and Q is the probability of unsuccessful operation (i.e., failure). Q is typically represented as P_F for probability of failure.

The following equations are used extensively in reliability and FTA quantitative calculations:

- $R + Q = 1$
- $R = P_S = e^{-\lambda T}$
- $Q = P_F = 1 - e^{-\lambda T}$
- $\lambda = 1 / MTBF$ or $\lambda = 1 / MTTF$
- Where λ is the failure rate and T is the exposure time.

Mean time between failures (MTBF) is a basic measure of reliability for *repairable items*. It is the mean number of life units during which all parts of the item perform within their specified limits, during a particular measurement interval under stated conditions. MTBF is the predicted elapsed time between inherent failures of a system, component or product during operation. MTBF can be calculated as the arithmetic mean (average) time between failures of an item. The MTBF is typically part of a model that assumes the failed item is immediately repaired (zero elapsed time), as a part of a renewal process. This is in contrast to the mean time to failure (MTTF), which measures average time between failures with the modeling assumption that the failed item is not repaired. Reliability increases as the MTBF increases. The MTBF is usually specified in hours, but can also be used with other units of measurement such as miles or cycles.

The equations for calculating MTBF are:

- MTBF = (total hours of operation) / (total number of failures)
- MTBF = $1 / \lambda$

Mean time to failure (MTTF) is a basic measure of reliability for *non-repairable items*. It is the total number of life units of an item population divided by the number of failures within that population, during a particular measurement interval under stated conditions. MTTF measures the average time between failures with the modeling assumption that the failed item is not repaired. Reliability increases as the MTTF increases. The MTTF is usually specified in hours, but can also be used with other units of measurement such as miles or cycles.

The equations for calculating MTTF are:

- MTTF = (total hours of operation) / (total number of failures)
- MTTF = $1 / \lambda$

Failure rate is an important measure in reliability engineering, system safety and FTA. Failure rate is the frequency with which a system, item or component fails. It is typically expressed in failures per hour or failures per million hours. It is usually denoted by the Greek letter lambda, using the "λ" symbol. Some industries use other frequency factors, such as failures per cycle or failures per mile.

14.3 Using Failure Data in FTA

In order to quantify a FT, the probability of failure value is required for all basic failure events in the FT. This value is typically derived from the failure rate for the item being investigated. Figure 14.1 demonstrates a FT quantitative evaluation using a small example FT with three system components. In FTA, failure rates are typically expressed in failures per hour, which is used in this example.

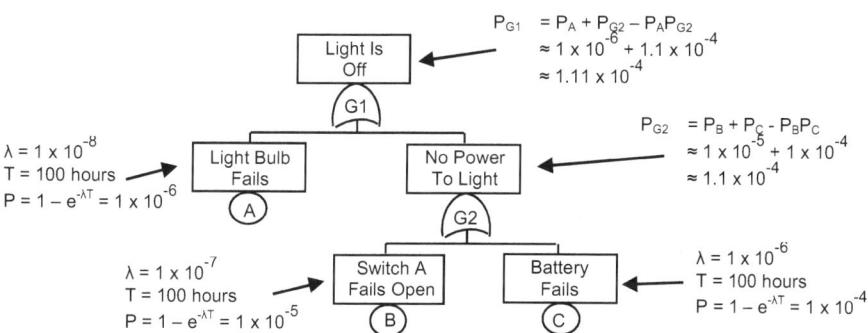

Figure 14.1 – Example FT Quantification

This example FT evaluation demonstrates how component failure rates are used in a quantitative FTA evaluation. The probability of failure of a component is a function of the component's failure rate and the component's exposure time during system operation. Probability is summed upward in the FT, based on the logic and probability equation for each gate in the FT.

14.4 Obtaining Failure Data

Where do failure rates come from? This is a key question and concern of FTA analysts. Failure rate data can be obtained in several ways, the most common sources include:

A. Historical data

Many organizations maintain internal databases of failure information on the devices or systems that they produce, which can be used to calculate failure rates for those devices or systems. For new devices or systems, the historical data for similar devices or systems can serve as a useful estimate.

B. Government and commercial data sources

Handbooks of failure rate data for various components are available from government and commercial sources. Mil-Hdbk-217F, *Reliability Prediction of Electronic Equipment*, is a military standard that provides failure rate data for many military electronic components. Several failure rate data sources are available commercially that focus on commercial components, including some non-electronic components. Many government agencies maintain databases of historical data on the systems they operate.

C. Testing

The most accurate source of data is to test samples of the actual devices or systems in order to generate failure data. Companies developing products typically perform testing to determine the reliability of their product. Testing items produced by manufacturers is a means for obtaining data, however, this method is often prohibitively expensive or impractical.

D. Research

Individual research is a feasible but often difficult method for obtaining failure data. Researching magazines and news articles for reported failures and operational hours of items is sometimes productive. Obtaining manufacturer's warranty claims for their products also produces claimed failure rates.

E. Analyses

Quite often existing analyses may provide the data needed. For example, a Failure mode and Effects Analysis (FMEA) generally contains failure data for components in the FMEA. If an FMEA has been performed on your system, it may be a good resource for failure data.

Commercial data sources are available that provide excellent data. Some common sources for failure data include the following:

A. Electronic Components
- MIL-HDBK-217, Reliability Prediction of Electronic Equipment
- Telcordia SR-332 (Bell Laboratories Bellcore)
- PRISM (Alion System Reliability Center (SRC))
- RIAC 217Plus (Reliability Information Analysis Center)

B. Mechanical Components
- NSWC Standard 98/LE1, Handbook of Reliability Prediction Procedures for Mechanical Equipment, U.S. Naval Surface Warfare Center, September 30, 1998.
- WASH-1400 Reactor Safety Study, 1975.

C. Human Error Rates
- Gertman, David I. & Blackman, Harold S., Human Reliability & Safety Analysis Data Handbook, John Wiley & Sons Inc., New York, NY, 1984.
- WASH-1400, Reactor Safety Study, 1975.
- THERP – Technique for Human Error Rate Prediction is a methodology used in the field of human reliability assessment (HRA), for the purposes of evaluating the probability of a human error occurring throughout the completion of a specific task.

Failure rates and failure modes for mechanical equipment types covered in NSWC Standard 98/LE1 includes the following:

- Actuators
- Bearings
- Brakes and Clutches
- Compressors
- Electric Motors
- Filters
- Gears and Splines
- Mechanical Couplings
- Miscellaneous Parts
- Pumps
- Seals and Gaskets
- Slider-Crank Mechanisms
- Solenoids
- Springs
- Threaded Fasteners
- Valve Assemblies

14.5 Failure Data Utility

FT analysts are typically concerned about the quality of FT failure data. FT quantitative results are only as accurate as the input data. Quite often it is difficult to obtain premium failure data with a proven history. When the data is questionable, FT reviewers often claim that the FT results are useless. Higher confidence data does provide higher confidence results; however, the old adage of "garbage in, garbage out" does not necessarily apply directly to quantitative FT results and data. When premium failure data is not available, worst case data can be used and still provide meaningful results. If the FT top probability is within acceptable ranges when worst case data is used, then it is not typically necessary to refine the data. If the top probability is not acceptable, then refine the data only for the system elements causing the problem.

The failure data used for FT quantification sometimes contains an element of uncertainty for various reasons, such as:

- No data is available for a component
- Data is obtained from partial historical records
- Data is obtained from limited sampling and testing
- Data is obtained from prediction method
- Data is obtained from manufacturer claims

Data uncertainty can be evaluated through the use of sensitivity analysis. The Risk Reduction Worth (RRW) and the Risk Achievement Worth (RAW) sensitivity measures help determine the quantitative effect of setting the probability of failure of a component to 0 and 1, respectively. If there is little effect on the top FT probability, then obtaining a more refined failure rate may be unnecessary, since it is not a major contributor to the top FT probability.

Remember, low confidence raw data can still provide useful results. If the top probability is acceptable when using worst case estimates, then more refined data may not be necessary. When the results are not desirable, then, find more accurate data for those items of concern only. In addition, sensitivity analysis can help determine if more refinement of the data is necessary.

CHAPTER 15

FTA HISTORY

In 1961, H. A. Watson and A. B. Mearns of Bell Laboratories conceived the FTA concept while performing a safety study of the Minuteman Launch Control System for the U.S. Air Force. The purpose of their study was to demonstrate a safe launch control system design; however, it evolved into a formal methodology for accomplishing design safety. Dave Haasl, then at the Boeing Company on the Minuteman program, recognized the value of FTA as an overall system safety tool; he led a team that applied FTA to the entire Minuteman Missile System. The Minuteman program used FTA to evaluate such UEs as "inadvertent programmed launch" and "inadvertent motor ignition" to quantitatively demonstrate that the design provided acceptable risk levels for these potential mishaps. Dave went on to lead a group of engineers and scientists in writing NUREG-0492, the first book on FTA and the unofficial FTA reference standard.

The commercial aircraft division of Boeing saw the results from the Minuteman program and quickly began using FTA during the design of commercial aircraft. In 1965, Boeing and the University of Washington sponsored the first System Safety Conference. At this conference, the first-ever papers were presented on FTA, marking the beginning of worldwide interest in the subject.

In 1966, Boeing developed a computer FT simulation program called BACSIM (Boeing Aerospace Corporation Simulation) for the evaluation of multi-phase fault trees. BACSIM could handle up to 12 operational phases, and included the capability for repair and K-factor adjustment of failure rates. The BACSIM program was developed by Bob Schroeder and Phyllis Nagel, both of Boeing. Bob Schroeder also developed a computer program that plotted FTs on a Calcomp 26-inch wide roll plotter. Both programs ran on an IBM 370 mainframe. These were specialized Boeing in-house programs, which few people were aware of outside the company.

Following the lead of the aerospace industry, the nuclear power industry discovered the virtues and benefits of FTA, and began using the tool in the design and development of nuclear power plants. Many key individuals in the nuclear power industry contributed to advancing FT theory and FT algorithms and computer programs. In fact, the nuclear power industry may have contributed more to the development of FTA than any other user group. Many new evaluation algorithms were developed, along with software implementing these algorithms, such as MOCUS, Prepp/Kitt, SETS, FTAP, Importance and COMCAN.

FTA has also been applied to the chemical process industry, the auto industry, rail transportation, launch vehicles and spacecraft, the aerospace industry and the robotics industry, just to name a few. There are probably many other industries and disciplines using FTA that have not been mentioned here. One of the more recent important events in the FTA methodology has been the development of commercial FT construction and evaluation software that operates on personal computers, which has provided great flexibility and utility for the safety analyst.

APPENDIX A
- REFERENCES -

Although there are many and references on FTA, the following books, documents and papers are some of the most relevant and useful. Strictly speaking, there are no official standards for FTA; however, the first three documents tend to serve as de facto standards.

1) NUREG-0492, Fault Tree Handbook, N. H. Roberts, W. E. Vesely, D. F. Haasl & F. F. Goldberg, original release 1981, 208 pages, U.S. Government Printing Office.

2) IEC-1025, Fault Tree Analysis (FTA), International Standard, International Electrotechnical Commission (IEC), First Edition, October 1990, 42 pages, English and French.

3) Fault Tree Handbook with Aerospace Applications, NASA, August 2002, version 1.1 (this is a revised and updated version of NUREG-0492).

4) Reliability and Risk Assessment, J. D. Andrews & T. R. Moss, Longman Scientific & Technical, 1993, 2nd edition 2002.

5) Probabilistic Risk Assessment and Management for Engineers and Scientists, E. J. Henley & H. Kumamoto, IEEE Press (2nd edition), 1996.

6) The Fault Tree Method, W. G. Schneeweiss, LiLoLe-Verlag, 1999.

7) Method for Obtaining Cutsets for Fault Trees, J. B. Fussell, W. E. Vesely, A New 1972, Transactions ANS, No. 15, p262-263.

8) MOCUS – A Computer Program to Obtain Minimal Cutsets, J. B. Fussell, et al, 1974, Aerojet nuclear ANCR-1156.

9) Dynamic Fault Tree Models For Fault Tolerant Computer Systems, J. Dugan, S. Bavuso and M. Boyd, IEEE Transactions on Reliability, Vol. 41, No. 3, September 1992, p363-377.

10) Dependency Modeling Using Fault Tree Analysis, J.D. Andrews & J.B. Dugan, Proceedings of the 17th International system safety Conference, 1999, p67-76.

11) Dependability Analysis of Systems With On-Demand and Active Failure Modes, Using Dynamic Fault Trees, L. Meshkat & J. B. Dugan & J. D. Andrews, IEEE Transactions on Reliability (Vol 51 No 2), 2002, p240-251.

12) Advanced Concepts in Fault Tree Analysis, Haasl, D. F., Proceedings of the Boeing and University of Washington System Safety Symposium, June, 1965.

13) Apollo Logic Diagram Analysis Guideline, Ericson, C. A. and Wolfe, B. A., Boeing document D2-117018-1, 1972.

14) System Safety Analytical Technology – Fault Tree Analysis, C. A. Ericson, Boeing document D2-113072-2, 1970.

15) Hazard Analysis Techniques for System Safety, Chapter 11 on FTA, pages 183-221, C. A. Ericson, Wiley, 2005.

16) Fuzzy Fault Tree Analysis Using Failure Possibility, G. Liang and M. Wang, Microelectronics and Reliability (Vol 33 No 4), 1993, pages 583-597.

17) Fault Tree Analysis Based on Fuzzy Logic, Li-Ping He, Hong-Zhong Huang and Ming Zuo, Proceedings of the Annual R & M Symposium; 2007, Pages 79-84.

18) Hybrid Fault Tree Analysis Using Fuzzy Sets, C. T. Lin and M. J. J. Wang, Reliability Engineering and System Safety (Vol 58 No 3), 1998, pages 205-213.

APPENDIX B
- EXAMPLE FAULT TREE ANALYSIS -

A.1 Problem Statement

The system in Figure A.1 represents a hypothetical aircraft electrical power system. This aircraft has two jet engines, each of which powers two electrical generators via bleed air from the engines. A minimum of two generators are required for aircraft electrical power. The system starts with generators G1 and G3 in the operating mode, with G3 and G4 acting as offline backup generators. When monitor M1 detects loss of electrical power from a generator, the computer turns on generator G3 and then G4 as necessary. If G1 fails, the system switches to G2 first, then G4 if G2 is failed. If G3 fails, the system switches to G4 first, then G2 if G4 is failed. Each generator also has internal fault monitoring data which it sends back to the computer, so that the computer can turn on the necessary backup generators. Keep in mind that this example system is relevant, but a little over-simplified in order to keep the amount of system detail to a minimum, thereby allowing the focus to be on the FT construction process.

The UE for this example FTA is "insufficient electrical power resulting from loss of power from 3 of 4 generators".

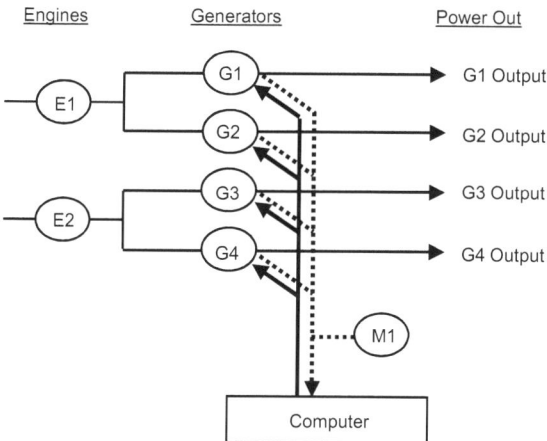

Figure A.1 – System Functional Diagram

A.2 FTA of Example Generator Problem

After reviewing several different FT approaches, it was decided to use a combined functional operation and functional flow scheme. First, the top level of the FT is first divided into the combinations of 3 out of 4 failures for the generators, i.e., functional operation. Then, the functional flow of each generator is analyzed backward. This involves starting at the power output point and working backwards towards the engines. This approach means looking at *all combinations* of three generators failing to provide output to the system (three out of four must be failed).

The FTs for this example are provided below. The wording in the text boxes is intentionally short and somewhat terse in order to keep the FT diagram small and simple. In an actual FTA more descriptive text would be provided.

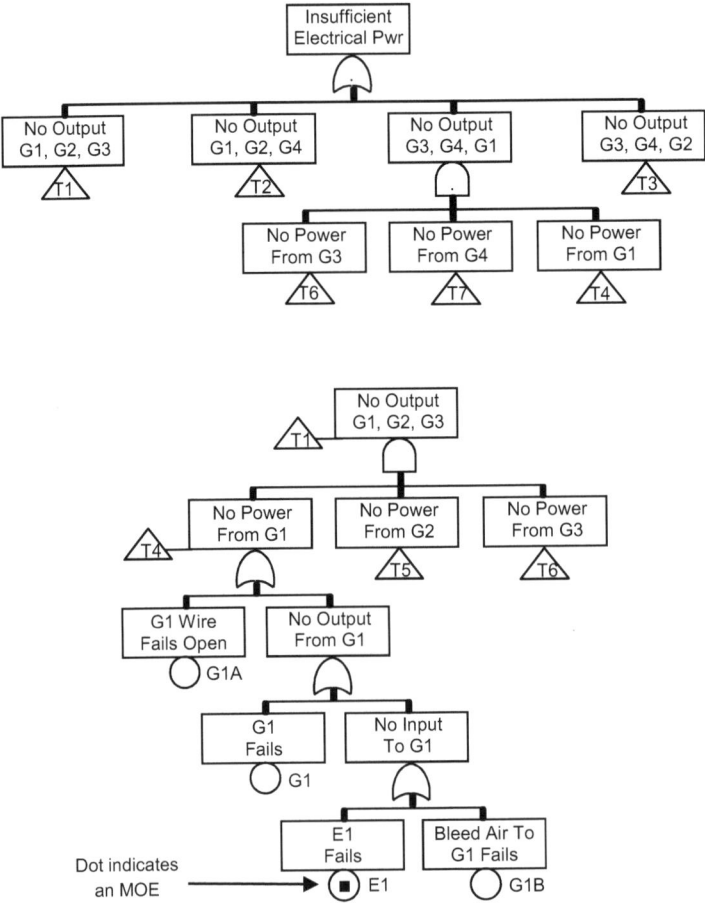

Fault Tree Analysis Primer

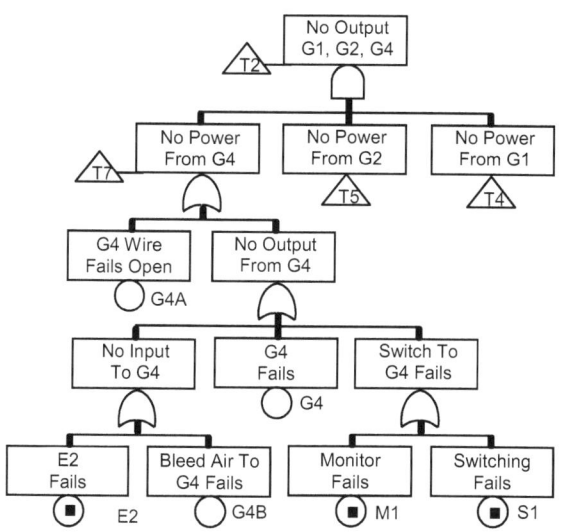

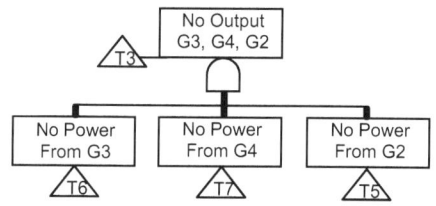

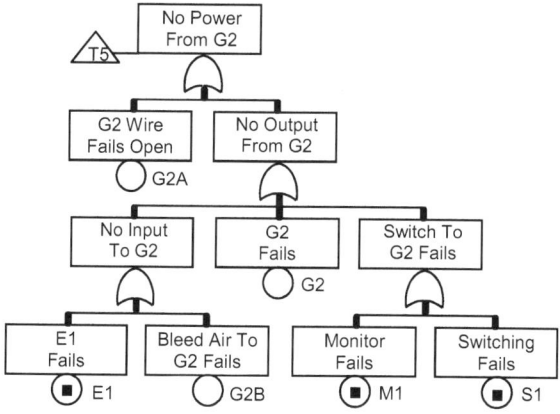

129

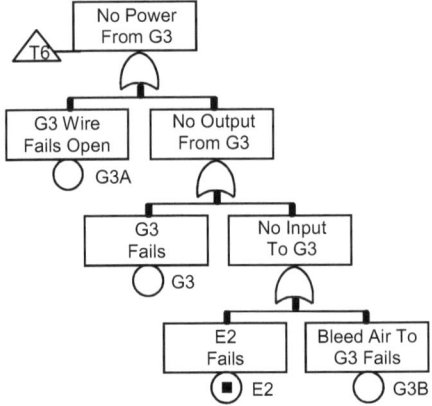

Things to note from the FT diagrams:

- All node names were limited to a maximum of 5 characters.
- A node naming convention was applied in order to easily distinguish subsystems, for example, events G1, G1A and G1b are related to generator 1.
- The FT structure followed the functional operation and then the functional flow approach.
- There are MOEs in the FT; indicated by a primary event symbol (circle) with a "■" symbol inside.

A FT computer program computed a total of 137 CSs for this FTA of the generator system. Each of these CSs represents one way the system can fail in the undesired mode. Of the 137 CSs, 29 were 2-order CSs and 108 were 3-order CSs. The CSs are listed in Table A.1.

Looking at the system design, one might at first assume that system failure requires a minimum of three combined failures (3-order CS), making the likelihood of failure very small. However, the MOEs in the system design create many 2-order CSs, which increase the probability of failure. Depending upon the design goals, a redesign might be necessary to eliminate the 2-order CSs.

This FTA demonstrates that a small number of system components can generate a large number of CSs. Each of these CSs represents a unique set of causal factors for potentially causing the UE to occur. Most system designers would likely not have guessed there were so many failure paths for this small system.

Table A.1 – CSs for Generator System FTA

137 Cut Sets from Generator FTA											
E1	E2		G1	G2A	G3	G1A	G2B	G3B	G1B	G3A	G4A
E1	G3		G1	G2A	G3A	G1A	G2B	G4	G1B	G3A	N47
E1	G3A		G1	G2A	G3B	G1A	G2B	G4A	G1B	G3B	G4
E1	G3B		G1	G2A	G4	G1A	G2B	N47	G1B	G3B	G4A
E1	G4		G1	G2A	G4A	G1A	G3	G4	G1B	G3B	N47
E1	G4A		G1	G2A	N47	G1A	G3	G4A	G2	G3	G4
E1	M1		G1	G2B	G3	G1A	G3	N47	G2	G3	G4A
E1	N47		G1	G2B	G3A	G1A	G3A	G4	G2	G3	N47
E1	S1		G1	G2B	G3B	G1A	G3A	G4A	G2	G3A	G4
E2	G1		G1	G2B	G4	G1A	G3A	N47	G2	G3A	G4A
E2	G1A		G1	G2B	G4A	G1A	G3B	G4	G2	G3A	N47
E2	G1B		G1	G2B	N47	G1A	G3B	G4A	G2	G3B	G4
E2	G2		G1	G3	G4	G1A	G3B	N47	G2	G3B	G4A
E2	G2A		G1	G3	G4A	G1B	G2	G3	G2	G3B	N47
E2	G2B		G1	G3	N47	G1B	G2	G3A	G2A	G3	G4
E2	M1		G1	G3A	G4	G1B	G2	G3B	G2A	G3	G4A
E2	S1		G1	G3A	G4A	G1B	G2	G4	G2A	G3	N47
G1	M1		G1	G3A	N47	G1B	G2	G4A	G2A	G3A	G4
G1	S1		G1	G3B	G4	G1B	G2	N47	G2A	G3A	G4A
G1A	M1		G1	G3B	G4A	G1B	G2A	G3	G2A	G3A	N47
G1A	S1		G1	G3B	N47	G1B	G2A	G3A	G2A	G3B	G4
G1B	M1		G1A	G2	G3	G1B	G2A	G3B	G2A	G3B	G4A
G1B	S1		G1A	G2	G3A	G1B	G2A	G4	G2A	G3B	N47
G3	M1		G1A	G2	G3B	G1B	G2A	G4A	G2B	G3	G4
G3	S1		G1A	G2	G4	G1B	G2A	N47	G2B	G3	G4A
G3A	M1		G1A	G2	G4A	G1B	G2B	G3	G2B	G3	N47
G3A	S1		G1A	G2	N47	G1B	G2B	G3A	G2B	G3A	G4
G3B	M1		G1A	G2A	G3	G1B	G2B	G3B	G2B	G3A	G4A
G3B	S1		G1A	G2A	G3A	G1B	G2B	G4	G2B	G3A	N47
G1	G2	G3	G1A	G2A	G3B	G1B	G2B	G4A	G2B	G3B	G4
G1	G2	G3A	G1A	G2A	G4	G1B	G2B	N47	G2B	G3B	G4A
G1	G2	G3B	G1A	G2A	G4A	G1B	G3	G4	G2B	G3B	N47
G1	G2	G4	G1A	G2A	N47	G1B	G3	G4A			
G1	G2	G4A	G1A	G2B	G3	G1B	G3	N47			
G1	G2	N47	G1A	G2B	G3A	G1B	G3A	G4			

Note: A space between CS elements represents an AND gate. For example, "E1 E2" means two items ANDed together, E1 AND E2.

APPENDIX C
- ABOUT THE AUTHOR -

Mr. Ericson has over 45 years of experience in the field of system safety, software design, software safety and Fault Tree Analysis (FTA). He holds a BSEE from the University of Washington and an MBA from Seattle University. Currently he works for the URS Corporation (formerly EG&G Technical Services) in Dahlgren, VA. He provides technical analysis, consulting, oversight and training on system safety and software safety projects. He currently supports NAVAIR system safety on the UCAS and BAMS unmanned aircraft systems, and he is assisting in writing NAVAIR system safety policies and procedures. Prior to joining URS, Mr. Ericson worked at Applied Ordnance Technology (AOT), Inc. of Waldorf, Maryland, where he was a program manager of system and software safety. In this capacity he directed projects in system safety and software safety engineering.

Prior to joining AOT, Mr. Ericson was employed as a Senior Principal Engineer for the Boeing Company for 35 years. At Boeing he worked in the fields of system safety, reliability, software engineering and computer programming. Mr. Ericson has been involved in all aspects of system safety, including hazard analysis, FTA, software safety, safety certification, safety documentation, safety research, new business proposals and safety training. He has worked on a diversity of projects, such as the Minuteman Missile System, SRAM missile system, ALCM missile system, Morgantown People Mover system, 757/767 aircraft, B-1A bomber, AWACS system, Boeing BOECOM system, EPRI solar power system and the Apollo Technical Integration program.

Mr. Ericson has taught courses on software safety and FTA at the University of Washington. Mr. Ericson was President of the System Safety Society in 2001-2003, and served as Executive Vice President of the System Safety Society, and Co-Chairman of the 16th International System Safety Conference. He was the technical program chairman for the 1998 and 2005 International System Safety Conferences. He is the founder of the Puget Sound chapter (Seattle) of the System Safety Society. In 2000 he won the Apollo Award for safety consulting work on the International Space Station, and the Boeing Achievement Award for developing the Boeing FTA course. Mr. Ericson won the System Safety Society's Presidents Achievement Award in 1998, 1999 and 2004 for outstanding work in the system safety field.

Mr. Ericson has prepared and presented training courses in system safety, software safety and FTA in the U.S., Singapore and Australia and has presented numerous technical papers at safety conferences. Mr. Ericson has published many technical articles on system and software safety and is currently editor of the Journal of System Safety (JSS), a publication of the International System Safety Society.

Mr. Ericson is the author of the NAVSEA Weapon System Safety Guidelines Handbook. Books published by Mr. Ericson include:

1) Hazard Analysis Techniques for System Safety, July 2005, Wiley.

2) Concise Encyclopedia of System Safety: Definition of Terms and Concepts, July 2011, Wiley.

3) System Safety Primer, September 2011, CreateSpace.

Mr. Ericson can be reached through his website www.risk-logic.com.

APPENDIX D
- INDEX -

AND Gate	15, 59
Command Fault	19
Boolean Laws	57
Boolean Reduction	61
Common Cause Failure (CCF)	20, 94
Common Mode Failure (CMF)	20
Cut Set	21, 59
Cut Set Order	21, 60
Cut Set Truncation	21, 60
Decision Tree	112
Event Tree	107
Exclusive OR Gate	15, 59
Exposure Time	22
Failure Data	117
Failure Modes and Effects Analysis (FMEA)	5, 85
Fault	22
Fault Tree Analysis (FTA)	3
Fault Tree Analysis Construction	42
Fault Tree Analysis Rules	71
Fault Tree Analysis Process	29
Gate Definitions	15, 59
Gate Probability Formulas	59
Inclusion-Exclusion Approximation	65
Importance Measure	24, 67
Inhibit Gate	15, 59
Interlock	96
Latency	93
Logic Tree	109
MinCS	21, 60
Min CS Upper Bound Approximation	66
Minimal Cut Set	21, 60
Mishap Tree	110
Multi-Phase FT	98
MOCUS Algorithm	63
Multiple Occurring Branch (MOB)	24
Multiple Occurring Event (MOE)	25
OR Gate	15, 59
Primary Failure	15, 25
Priority AND Gate	15, 59
Probability Laws	55
Probability of Loss of Aircraft (PLOA)	51
Probabilistic Risk Assessment (PRA)	25

Redundancy	101
Reliability	26, 57
Reliability Block Diagram (RBD)	26
Root Cause Analysis	37
Secondary Failure	15, 26
Sensitivity Measure	26, 67
Single Point Failure (SPF)	26
State of the Component	42
State of the System	42
Success Tree	107
System	27
System Safety	27
Top Undesired Event (TUE)	28
Transfer	17
Undesired Event UE)	28

Printed in Great Britain
by Amazon.co.uk, Ltd.,
Marston Gate.